P9-DVV-392

PLUMBING

Other Publications:

THE GOOD COOK
THE SEAFARERS
THE ENCYCLOPEDIA OF COLLECTIBLES
THE GREAT CITIES
WORLD WAR II
THE WORLD'S WILD PLACES
THE TIME-LIFE LIBRARY OF BOATING
HUMAN BEHAVIOR
THE ART OF SEWING
THE OLD WEST
THE EMERGENCE OF MAN
THE AMERICAN WILDERNESS
THE TIME-LIFE ENCYCLOPEDIA OF GARDENING
LIFE LIBRARY OF PHOTOGRAPHY
THIS FABULOUS CENTURY
FOODS OF THE WORLD
TIME-LIFE LIBRARY OF AMERICA
TIME-LIFE LIBRARY OF ART
GREAT AGES OF MAN
LIFE SCIENCE LIBRARY
THE LIFE HISTORY OF THE UNITED STATES
TIME READING PROGRAM
LIFE NATURE LIBRARY
LIFE WORLD LIBRARY
FAMILY LIBRARY:
HOW THINGS WORK IN YOUR HOME
THE TIME-LIFE BOOK OF THE FAMILY CAR
THE TIME-LIFE FAMILY LEGAL GUIDE
THE TIME-LIFE BOOK OF FAMILY FINANCE

HOME REPAIR
AND IMPROVEMENT

PLUMBING

BY THE EDITORS OF
TIME-LIFE BOOKS

TIME-LIFE BOOKS
ALEXANDRIA, VIRGINIA

Time-Life Books Inc.
is a wholly owned subsidiary of
TIME INCORPORATED

Founder Henry R. Luce 1898-1967

Editor-in-Chief Henry Anatole Grunwald
Chairman of the Board Andrew Heiskell
President James R. Shepley
Editorial Director Ralph Graves
Vice Chairman Arthur Temple

TIME-LIFE BOOKS INC.

Managing Editor Jerry Korn
Executive Editor David Maness
Assistant Managing Editors Dale M. Brown (planning), George Constable, George G. Daniels (acting), Martin Mann
Art Director Tom Suzuki
Chief of Research David L. Harrison
Director of Photography Robert G. Mason
Senior Text Editors William Frankel, Diana Hirsh
Assistant Art Director Arnold C. Holeywell
Assistant Chief of Research Carolyn L. Sackett
Assistant Director of Photography Dolores A. Littles

Chairman Joan D. Manley
President John D. McSweeney
Executive Vice Presidents Carl G. Jaeger, John Steven Maxwell, David J. Walsh
Vice Presidents Nicholas Benton (public relations), John L. Canova (sales), Nicholas J. C. Ingleton (Asia), James L. Mercer (Europe/South Pacific), Herbert Sorkin (production), Paul R. Stewart (promotion), Peter G. Barnes
Personnel Director Beatrice T. Dobie
Consumer Affairs Director Carol Flaumenhaft
Comptroller George Artandi

HOME REPAIR AND IMPROVEMENT

Editorial Staff for Plumbing
Editor John Paul Porter
Assistant Editor Edward Brash
Picture Editor Kaye Neil Noble
Designer Herbert H. Quarmby
Associate Designer Robert McKee
Text Editors Anne Horan, Gerry Schremp
Staff Writers Richard Bayan, Barbara Ensrud, Sally French, Angela D. Goodman, Simone Gossner, Lee Greene, Kumait Jawdat, Michael Luftman, Joan Mebane, Don Nelson
Researchers: Ginger Seippel, Scot Terrell
Art Associates: Angela Alleyne, Faye H. Eng, Kaye Sherry Hirsh, Richard Salcer, Victoria Vebell, Mary Wilshire
Editorial Assistant: Eleanor G. Kask

Editorial Production
Production Editor Douglas B. Graham
Operations Manager Gennaro C. Esposito, Gordon E. Buck (assistant)
Assistant Production Editor Feliciano Madrid
Quality Control Robert L. Young (director), James J. Cox (assistant), Michael G. Wight (associate)
Art Coordinator Anne B. Landry
Copy Staff Susan B. Galloway (chief), Ricki Tarlow, Eleanor Van Bellingham, Florence Keith, Celia Beattie
Picture Department Barbara S. Simon
Traffic Jeanne Potter

Correspondents: Elisabeth Kraemer (Bonn); Margot Hapgood, Dorothy Bacon, Lesley Coleman (London); Susan Jonas, Lucy T. Voulgaris (New York); Maria Vincenza Aloisi, Josephine du Brusle (Paris); Ann Natanson (Rome). Valuable assistance was also given by Carolyn T. Chubet, Miriam Hsia (New York).

THE CONSULTANTS: Thomas P. Konen, the general consultant for this book, is chief of the Building Technology Research Division, Stevens Institute of Technology, Hoboken, New Jersey.

Harris Mitchell, special consultant for Canada, has been working in the field of home repair and improvement for more than two decades. His experience ranges from editing *Canadian Homes* magazine to writing a syndicated newspaper column, "You Wanted to Know," and he is the editor or author of a number of books on home improvement.

Gerard Drohan, a licensed master plumber, is a specialist in alteration and repair work. He also teaches courses in basic plumbing at the Mechanics Institute in New York City.

© 1979, 1977, 1976 Time-Life Books Inc. All rights reserved.
No part of this book may be reproduced in any form or by any electronic or mechanical means, including information storage and retrieval devices or systems, without prior written permission from the publisher, except that brief passages may be quoted for reviews.
Fourth printing.
Published simultaneously in Canada.
Library of Congress catalogue card number 76-46139.
School and library distribution by Silver Burdett Company, Morristown, New Jersey.

Contents

1 New Scope for the Home Plumber

Pipes in your home. The network of piping at left includes the three main types—drain, vent and supply—of a plumbing system. The thickest *(second from left)* is the main drain stack, which channels all wastes to a sewer line or septic tank; other waste pipes—such as the branch drain at lower right—empty into the main stack. To keep air moving through the drains and to eliminate sewer gases, vent pipes *(far left)* connect drains to a roof opening. The thin vertical and horizontal pipes fitted with shutoff valves are supply lines carrying hot and cold water to individual fixtures.

Most people fix dripping faucets as a matter of routine. Beyond that chore, plumbing generally has been left to plumbers. Yet today a whole list of tasks once the province of professionals—replacing old fixtures and faucets with modern easier-to-use types, extending pipes to add or relocate equipment, installing purification filters, special valves or built-in sprinklers—can be and often must be handled by amateurs. Economy is one reason; the professional must charge so much for his time that a householder may save money completing a job himself even if he buys a special tool to be used only once. Perhaps more important is the revolution in plumbing techniques that has eliminated the requirements for brute strength and dexterous skills; new materials are light in weight and many are assembled simply by tightening nuts or applying glue. And an increasingly significant factor in the trend to do-it-yourself plumbing is the growing concern for a resource long taken for granted. Water is constantly getting more expensive: care and economy in its use in private plumbing systems has become essential.

Every family can keep its water bills down—while enjoying the many conveniences of new equipment—by a regular program of repairs and minor alterations.

☐ Repairing a drip *(pages 32-35)* while it is still in the drop-by-drop stage saves an amazing quantity of water—a trickle wastes a bathtubful of water a day, and a steady stream wastes enough water to meet all of a family's daily needs.

☐ Inexpensive aerators, easily fitted into sink and lavatory spouts *(pages 44-45)*, deliver a splash-free stream and pay for themselves many times over in water savings.

☐ Replacing an existing toilet *(pages 98-99)* with one of the new water-saving models can reduce water consumption by gallons a day —and you can keep the saving at a maximum by a simple adjustment that controls the length of the flush cycle *(pages 62-63)*.

☐ A dry well *(pages 106-109)*—a separate, simple drainage system for certain appliances—lightens the burden on a septic tank so that it may not need costly cleaning so often.

Using the right tools and methods, you can meet the crises and increase the conveniences of your plumbing system—seal a burst pipe or thaw a frozen one *(pages 18-21)*, unclog a stopped-up main drain *(pages 26-27)*. The modern plumbing materials and fittings that have made these tasks simpler have been largely standardized; the pipes, fittings and fixtures described in this book are available everywhere, from the Gulf of Mexico to the provinces of Canada. Some varieties of copper tubing and the increasingly popular plastic pipe are flexible and easy to install. To attach a connection to these pipes, you may not need even to solder or cement a fitting: in many situations a compression-type fitting or a simple mechanical tap-on device *(page 73)* will do the job.

You will be able to use these new repair and modernization methods with confidence if you understand how the plumbing system operates. Some of the physical principles are intriguing aspects of fundamental science *(pages 16-16H)*, but most that you need to know about are very simple. In a home plumbing system, cold water from public mains or your own well flows into the house through a service pipe. Inside the house, branch supply lines bring the water under pressure to all plumbing fixtures and most appliances. A separate supply line carries cold water to a water heater; from there, a network of pipes supplies hot water under pressure to the fixtures and appliances that use it. In this supply system the key element is pressure—40 pounds per square inch in most localities, even higher in some large cities.

While the entire supply system brings in water by pressure, the drainage system carries off waste water and sewage by the force of gravity. Drainpipes always slant downward or run vertically, and bends are gentle to make the wastes flow smoothly. The crucial element in the drain system, comparable to the main service line of the supply system, is the large vertical drain called the main soil stack. All toilets drain directly into the main stack; other fixtures and appliances may empty into slanting branch drains, which connect to the main stack. It, in turn, empties into a slanting main drain—generally laid just beneath the basement floor—that carries the wastes to a public sewer or a septic tank.

At several points in this system are elements intended to keep air passing through the sewer and drains but not into the house. The free passage of air is essential to maintain atmospheric pressure at every point; otherwise a vacuum will develop that will stop waste water from flowing down. This air should not enter living space, of course, for it contains sewage gases.

To keep the sewage gases out of the rooms of a house, every fixture and some drain lines empty into traps—U-shaped passages, permanently filled with water that serves as a barrier to the gases. Sink and lavatory traps are generally accessible and easy to clear or, if necessary, to replace; bathtubs, shower stalls and toilets, which empty their wastes through the floor, can be cleared, but replacing the traps alone is difficult.

While traps seal air inside the drains, it passes through them freely. When waste water flows, air goes downward with it; when no waste flows, air passage reverses and rises in the pipes to escape from the house. This two-way movement is made possible by a venting system—a network of pipes linking many branch drains to the main soil stack, which has a chimney-like pipe rising through the roof to let air in and out.

Alterations of any kind in your plumbing system are controlled by local regulations in your community. There is no nationwide code for the United States or Canada—the so-called National Plumbing Code of the United States sets minimum standards only for federally built structures and may not apply to your home. Local codes have the force of law, and must be observed, but none of them prevents you from working on your own plumbing system, so long as you follow the provisions in the codes. Because they are established to meet the special problems of the particular area in which you live, observing them is not only necessary, but is also a wise precaution for your own health and safety.

Depending on the soil of your locality, your water may be hard or soft—that is, relatively rich or poor in dissolved minerals. In turn, the nature of your water affects the kind of pipe that you should use. Almost everywhere in the United States and Canada, for example, copper pipe is acceptable—but not quite everywhere. In some areas of hard water, minerals build deposits on the inner walls of copper tubing so quickly that supply lines clog. In some soft-water areas, on the other hand, copper tubing is vulnerable to acids in the water. In parts of both areas, copper pipe is prohibited: plumbers must use plastic or galvanized steel.

Venting, too, is affected by local conditions. Normally, a main vent consists of a single pipe of a standard size rising above the roof line. But in the harsh winters of some parts of Canada and the United States, snow and ice would soon block the standard vent; codes there call for outsize vent pipes.

You can consult your code at the office of your local health department or building inspector; if you plan extensive work, it is helpful to have a copy of your own. You will not need it for an ordinary repair or replacement, but in a new installation it is an indispensable guide to the right plumbing materials and the right way to put them together.

Tool Kit for Plumbing

With only a few specialized tools you can be ready not only to meet most plumbing emergencies, but also to install and replace pipes and fixtures.

□ For unclogging drains you will need a plunger—the "plumber's helper." A cup extension makes the plunger shown effective for toilets; it can be folded inside for flat drains. A trap-and-drain auger or snake draws out blockages from sink drains; for toilets, use a closet auger, housed in a cylinder that positions the hooked end.

□ To loosen and tighten joints and fixtures several specialized wrenches are helpful. A pipe wrench bites into and holds onto pipe as you turn it. Wrenches for supply pipe should measure 12 to 14 inches from end to end, those for waste pipe 18 inches. A basin wrench tightens nuts behind sinks and in other tight spaces. The spud wrench has wide, toothless jaws for the large nuts on toilets and sinks. A torque wrench secures sections of cast-iron pipe.

□ To fix leaks caused by worn valve seats *(page 34)* you need a seat wrench, which has a square tip on one end and an octagonal tip on the other to grip either of the two common types of seats so that you can replace them. For older faucets with nonreplaceable seats you need a seat dresser with assorted cutting disks and guides to grind smooth the nicks.

□ For cutting and joining copper tubing, use a tube cutter with a built-in reamer for scraping off burrs around the cut edge. A coiled tube bender allows you to bend tubing without kinks. To join sections of copper tubing or pipe by sweat soldering, the materials required are emery cloth, flux and a flux applicator, a propane torch, solder and asbestos board or sheeting to protect nearby framing. The flame-spreader attachment helps thaw frozen pipe. A flaring tool spreads the ends of tubing for flared joints.

□ For cutting cast-iron pipe you may want to purchase a special tool that has steel disks to get through pipes up to 4 inches in diameter. For pipes over 4 inches, larger cutters can be rented. Working with plastic pipe calls for no special tools, only glue and solvent *(page 77)*.

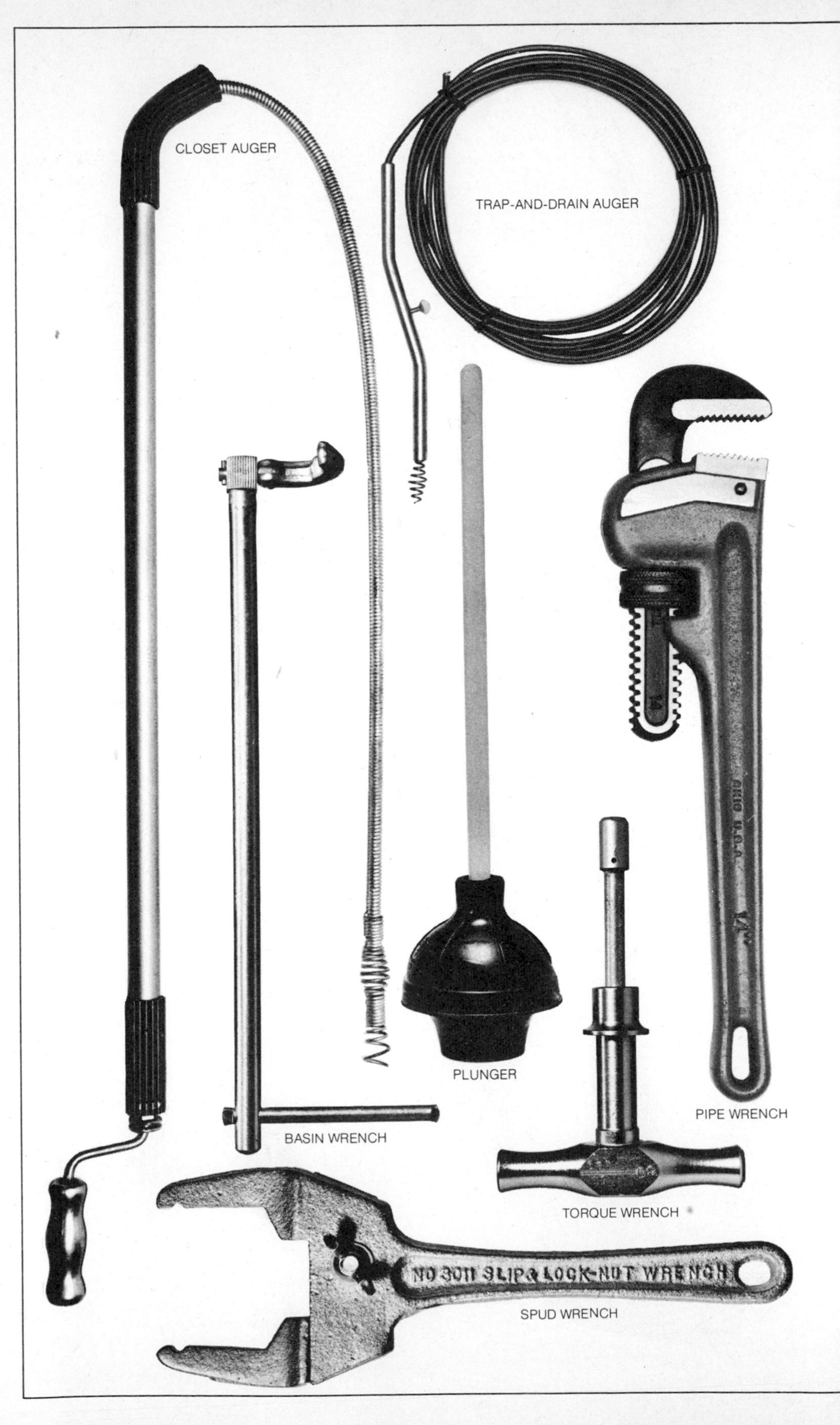

FLAME-SPREADER ATTACHMENT
ASBESTOS
SEAT WRENCH
SEAT DRESSER
TUBE CUTTER
SOLDER
FLUX
FLUX APPLICATOR
PROPANE TORCH
TUBE-FLARING TOOL
EMERY CLOTH
TUBE BENDER

The Meter: A Record of Water Used or Wasted

The water meter in your house is put there to register the amount of water used during the utility company billing period. As far as the company is concerned, bill-gauging is the sole function of the meter, but you can use it for other purposes—to find out how much of your water bill is run up by certain jobs, such as laundry or lawn-sprinkling, or even to help discover a leak—provided you know how to read the meter correctly.

There are two types of meter face. The direct-reading type (not illustrated) gives a total at a glance, like a car odometer. The cumulative-reading meter has a number of dials that must be read separately, then combined *(below)* to get total usage in cubic feet. Regardless of type, all meters have a special pointer that makes a complete revolution for each cubic foot of water being consumed at a given moment.

To measure the water you use when sprinkling the lawn, examine the meter before turning on the sprinkler and again after turning it off. The difference between the two figures is the number of cubic feet of water you have used on the lawn. To convert the answer to gallons, multiply it by 7.5, the number of gallons in one cubic foot. (While making such a test you should be sure that there are no other heavy demands on your plumbing system.) The same method works, of course, if you want to determine water usage for anything from washing a car to filling a swimming pool.

If you want to find out if you have a leak, the fast-moving pointer is the one to watch. Suppose, for example, you have a fresh-water stain on a ceiling and suspect a hidden pipe leak is the cause. First be sure that all faucets are turned off, then watch the pointer for a few minutes. If the pointer advances, however slowly, there is a leak in your plumbing. If it stands still, the stain is due to another cause, such as a faulty roof.

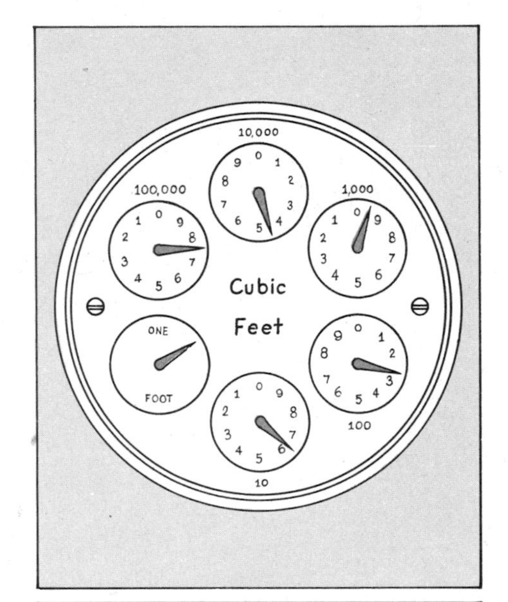

Reading a six-dial meter. Start at the dial labeled 100,000 and note the smaller of the two digits nearest the pointer (bear in mind that the pointers move alternately counterclockwise and clockwise from one dial to the next). In the same way read the dials labeled 10,000, 1,000, 100 and 10 in that order. The five digits you read off each dial, written consecutively, provide the current reading on your meter, expressed in cubic feet (74,926 in the case pictured). The sixth dial, labeled "one foot," is used only to detect leaks.

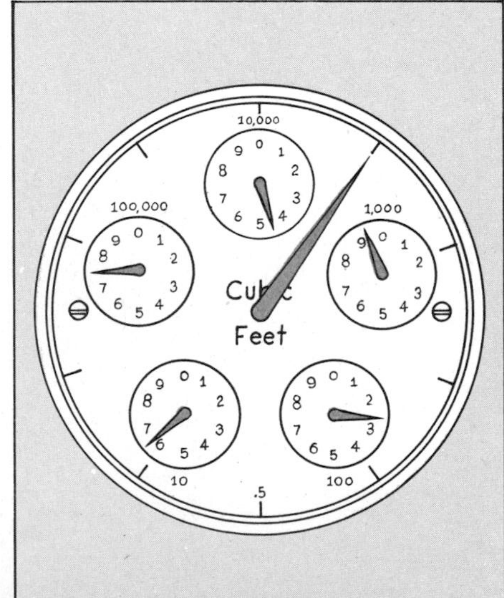

Reading a five-dial meter. The major difference between this meter and the one above is that the one-cubic-foot flow is indicated by a long pointer that sweeps the outer edge of the meter face instead of by a small individual dial. The other difference is that all the pointers on the small faces move clockwise. Otherwise the meter is read in the same way: Note the dial figures one by one, starting with the 100,000 dial. The cumulative reading here is also 74,926.

A Gurgle Detector for Finding Leaks

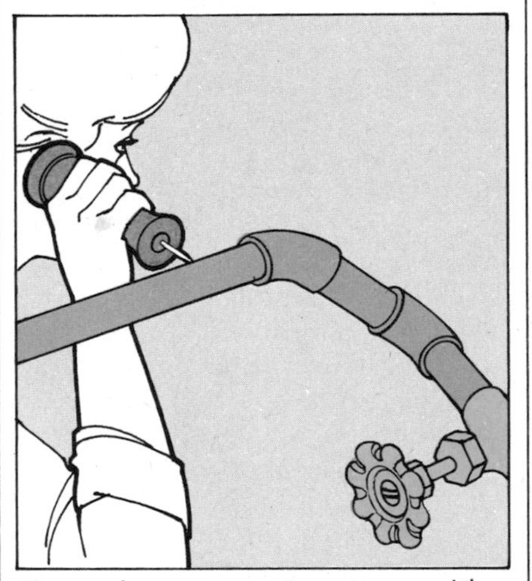

If you have no water meter—either because you have your own well or because you live in a community where residential water meters are not required—you can verify the existence of a hidden leak with any of several sound-amplifying devices available from plumbing-supply stores. The one above looks like an old-fashioned telephone receiver but has only mechanical parts: a rod to pick up vibrations and a disk at the listening end to amplify these vibrations. To use it, check that all faucets are turned off, then hold the point against the main supply pipe. Put your ear to the wide end and listen. If water is moving through the pipe—and leaking from it—you will hear a gurgle.

Proper Care for a Water Heater

Few modern conveniences are more taken for granted than hot water. And most of the time, the tank that provides it can be taken for granted too. But there are two routine chores that must be done from time to time if you want it to do its job efficiently—and safely. One is clearing the tank of sediment; the other is checking the tank's relief valve.

Sediment—the result of rusty or alkaline impurities in many areas—comes into the hot-water tank from the water main. If it accumulates inside the tank, it blocks the transmission of heat to the water and wastes energy. But sediment settles near the bottom of the tank and is easily drawn off through a drain valve *(right)*. How often the tank needs draining depends on the composition of the water and the condition of the tank. Experiment by checking every month until you establish a cycle that allows no more than a pailful of cloudy water to accumulate between drawings.

The other maintenance chore—checking the water heater's relief valve—is a safety measure. The valve is designed to backstop the thermostat. In the unlikely—but not impossible—event that the thermostat should malfunction and permit the temperature to rise to a dangerous level, the valve will open automatically and release the overheated water before it can boil into steam and cause an explosion. To be sure that the valve is in working condition, make the simple test shown at right.

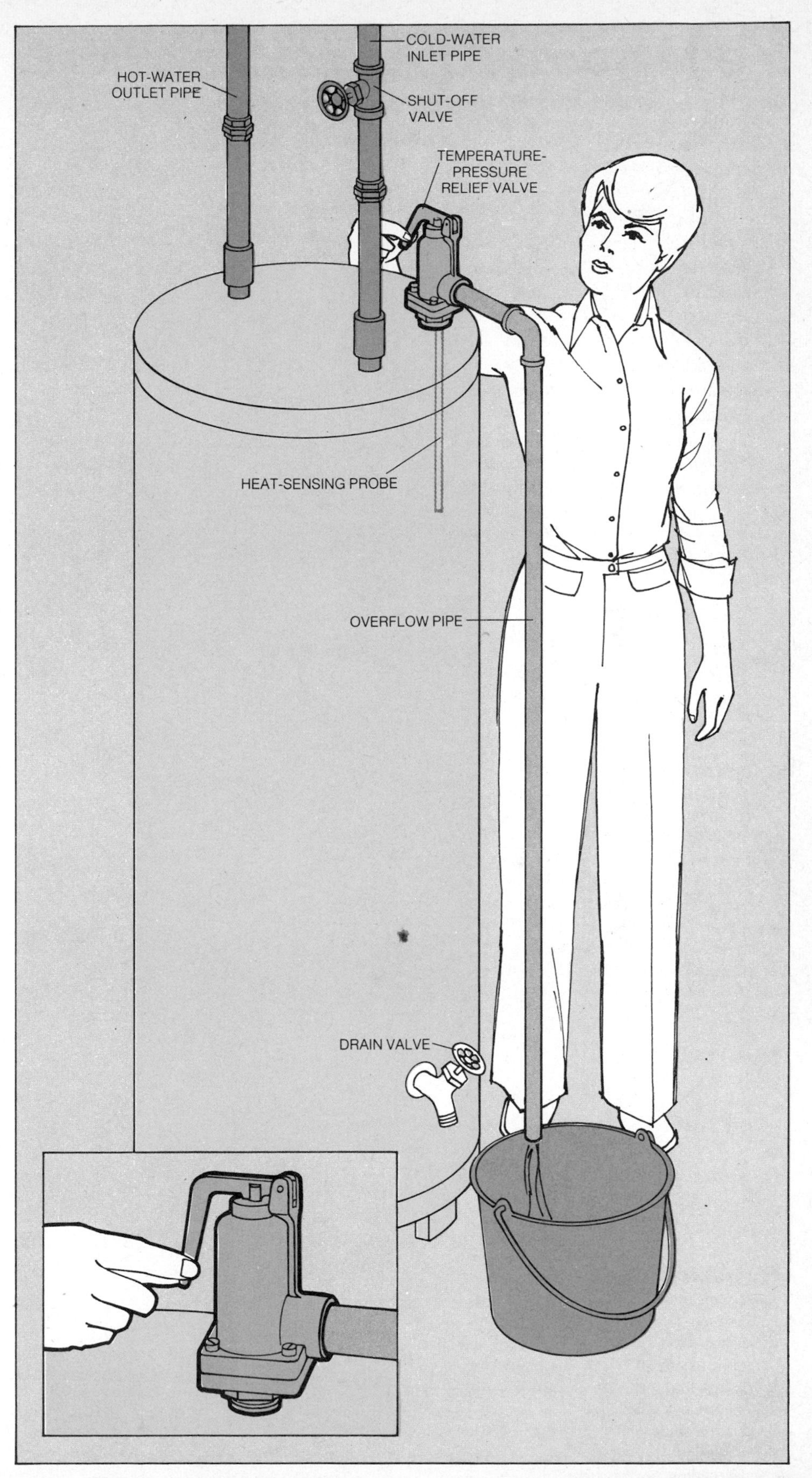

Checking a water heater. In a typical heater, the safety valve is on top, controlled by a probe beneath it, inside the tank. The probe automatically trips the valve if the temperature or pressure inside the heater exceeds safe limits. Once a season, lift the handle atop the valve *(inset)*. The overflow pipe below it should expel hot water; if it does not, replace the valve at once *(page 103)*.

The valve in the pipe above the tank controls the cold-water supply. At the bottom of the tank is the drain valve, to check for sediment. If water is cloudy or rusty, drain until it runs clear.

Keeping Your Water Supply Clean and Fresh

Perfectly pure water does not exist. Even as it condenses in a rain cloud and falls to earth, water absorbs carbon dioxide gas from the atmosphere and picks up minute particles of soot and dust. Other foreign ingredients are added after the rain begins to percolate through the soil. The water gathers up particles of dirt, dissolves a variety of minerals, creates new compounds by chemical reaction with soil and rock and becomes the home of colonies of bacteria. By the time the water completes its circuitous journey to your tap, most of the impurities should have been removed at the local water-treatment plant, but unfortunately such is not always the case. In addition, over 40 million Americans depend on their own wells for their water needs, and thus must provide their own equipment to remove any impurities, including:

☐ HARD WATER. This problem—the most common one of all—occurs to some degree in approximately 80 per cent of the water in the United States and Canada and is almost as common in municipal water as it is in well water. It is caused mainly by dissolved compounds of calcium and magnesium, which form a scaly build-up in water pipes, washing machines, water heaters and faucets, and can eventually block them and reduce their efficiency. The minerals also interfere with the action of most hand soaps, some shampoos and, to some extent, even modern laundry detergents.

☐ IRON. Dissolved iron compounds are most common in the well water of the Midwest and Florida although they can also be found in other areas. They leave rusty deposits on everything they touch —sinks, dishes, clothes washers and even the clothes themselves.

☐ CORROSIVE WATER. When substantial amounts of carbon dioxide are absorbed in water, chemical reactions create carbonic acid. This compound slowly eats away pipes, fittings and appliances. The condition is most common in the East and the Pacific Northwest.

☐ SULFUR WATER. Dissolved hydrogen sulfide gas gives water the unpleasant smell associated with rotten eggs. Sulfur water also tarnishes silver and can give tap water a blackish tinge. It can occur anywhere, but is most common in the Southwest and Midwest.

☐ CLOUDY OR DISCOLORED WATER. This problem occurs mostly in water drawn from lakes and rivers, or wells in heavily vegetated lowland areas. It is caused by suspended particles—either silt, algae or organic material—which affect the taste and appearance of the water and can also clog or damage valves, washing-machine parts and faucets.

☐ CONTAMINATION. Water that is polluted with disease-producing bacteria or viruses is not common in either well water or utility-supplied water—but it can occur. Despite federal, state and local laws, treatment plants do not always remove enough germs from water, and testing is sometimes inadequate. The Environmental Protection Agency has estimated that as many as 5,000 community water supply systems provide water that is potentially dangerous to drink. Water from private wells is even more susceptible to accidental contamination.

Pinpointing Potential Problems

The table opposite describes clues that will help you to define the nature of problems in your water supply. If these crude checks prove positive, have your water tested by a professional laboratory to see if additional water treatment is called for. Since there is no sure way for the homeowner to detect the presence of bacteria, a lab test should be made for any new well. An existing well should be tested every six months if there is a known contamination problem in the area. Your local board of health will do a bacterial analysis at no charge and can recommend a commercial laboratory that will examine the water for dissolved minerals, acid or organic matter. The extension service of the federal or state Department of Agriculture may also do free tests, as will companies that sell home water-purification equipment.

Conditioning Equipment for the Home

A wide range of home water-treatment equipment is available, and some can solve several problems at once; each device can be used with others to cure almost any combination of conditions. Not surprisingly, one of the best-selling water-treatment devices is a water softener: this is simply a tank containing an "ion-exchange resin." As the water passes through the softener, calcium and magnesium in dissolved compounds are exchanged for the sodium in the resin and remain with the resin in the tank. When the sodium in the resin is depleted, the tank must be recharged by back-flushing it with a salt solution—sodium from the salt is then exchanged for the calcium and magnesium, which empty into the drain. The simplest setup is a single tank owned by a private water-treatment company that periodically replaces it with a freshly charged tank. But some homeowners opt for a permanent installation. It includes, in addition to the softener tank, a tank filled with salt crystals, a pump and a sensing device or timer that automatically back-flushes and recharges the softener when the sodium in its resin is depleted.

A water softener will also remove some iron materials and suspended particles from the water by trapping them in its resin bed. Its only drawback is the high sodium content of the water it softens: persons on a low-salt diet because of heart or kidney trouble should not drink water processed by such a device, and some authorities have questioned the consumption of large amounts of sodium by anyone. For this reason, many people connect a softener only to the hot-water line—where minerals precipitate at a greater rate—leaving untreated the water used for drinking and cooking.

The most versatile water-treatment device is the chemical feeder that is installed on many private well systems. It is a container with an automatic pump that injects a small, precisely metered amount of a chemical solution into the supply line. The feeder can be used to cure a wide variety of problems, depending on the chemical used. To remove bacteria or large amounts of hydrogen sulfide, for example, the feeder is filled with a chlorine mixture. An alkaline solution is used to neutralize acid, and other chemicals can be injected to eliminate suspended particles.

Filters, the simplest water-treatment devices, can be used either alone or in conjunction with a water softener. When used with a softener, a large filter is usually installed in the line ahead of the water softener to screen out particles that might otherwise clog it; a small filter, like the charcoal-core type on page 100, is placed near the outlet to be filtered.

Common filter units include sand filters, neutralizing filters and oxidizing filters. The sand filter simply traps suspended particles of silt, algae and organic material. Neutralizing filters contain marble or limestone chips that react with the carbonic acid and weaken it, but they are not as effective as a chemical feeder for strongly acidic water. Oxidizing filters use manganese-treated sand to remove larger amounts of iron from water than a water softener can handle, and are also effective in reducing hydrogen sulfide.

Each of these filters must be backflushed frequently to remove trapped particles. Neutralizing and oxidizing filters must also be replenished a couple of times a year with new core material, while the core of a charcoal filter must be replaced at least every six months so that any bacteria in the water cannot feed and multiply on the organic materials that the filter traps.

The most sophisticated treatment device is the reverse-osmosis purifier. This unit removes hardness minerals, salt and suspended particles in one process by forcing the water under pressure into a thin membrane through which only pure water can pass. The unwanted materials remain on one side of the membrane and are drained from the unit through a waste outlet. Its capacity is low, and it is attached under the sink or lavatory to provide drinking and cooking water from a single tap.

Common Water Problems—and Their Remedies

Problem	Cause	Solution
Soap sludge in lavatories and bath tubs after use. White scaly deposits in faucets, shower heads, pipes and water heaters.	Magnesium and calcium compounds	An ion-exchange water softener.
Rusty deposits in sinks, tubs and washing machines, and in washed fabrics. Water left standing turns reddish after exposure to air.	Iron compounds	For small amounts, an ion-exchange water softener. For larger amounts, an oxidizing filter.
Reddish stains below faucets in tubs and sinks served by steel pipes. Greenish stains in fixtures served by copper pipes.	Carbonic acid	For small amounts, a neutralizing filter. For larger amounts a chemical feeder with an alkaline solution.
"Rotten egg" smell. Tarnishes silver. Blackish tinge.	Hydrogen sulfide	For small amounts, an oxidizing filter. For larger amounts, a chemical feeder with a chlorine solution followed by sand filter.
Cloudy or dirty appearance.	Suspended particles of silt, mud or sand	A sand filter.
Yellow or brownish tinge, or unpleasant taste.	Algae or other organic substance suspended in water	For small amounts, a charcoal-core filter. For larger amounts, a chemical feeder with chlorine or alum solution, followed by a sand filter or, for potable water only, a reverse osmosis unit.
Bacteria indicated by laboratory analysis.	Well water contaminated by improperly disposed sewage nearby	Correct improper sewage disposal. Install a chemical feeder with strong chlorine solution, followed by a charcoal-core filter.

What Goes On inside the Pipes

A complete home plumbing system requires four distinct networks of pipes. Two of them, the hot- and cold-water supply lines, carry water under pressure as high as 45 pounds per square inch. A third network, the drain pipes, works entirely by gravity. And a fourth carries no water at all; instead, it serves to maintain equal air pressure throughout the drain pipes so that their flow proceeds without interference.

The color-coded model opposite indicates how complex these networks can be in even a modest-sized house. But stripping away the structure of a house to reveal its pipes and fittings can present only an incomplete picture of a plumbing system. Ultimately, both the virtues and the vulnerabilities of the system depend on the behavior of water in the various conduits.

In order to expose the hidden world within the pipes, the Editors of TIME-LIFE BOOKS collaborated with scientists at a leading plumbing research facility—the Davidson Laboratory of Stevens Institute of Technology in Hoboken, New Jersey—to develop a series of laboratory demonstrations using plastic tubing to represent home equipment. In these demonstrations, recorded in dramatic color photographs, can be seen the action of important plumbing phenomena, both good and bad.

If the Stevens Institute experiments illustrate a single overriding point, it is that nothing in a plumbing system should be taken for granted. The pipes and valves are among the sturdiest parts of a house; they should last for several lifetimes. But things go wrong, often in subtle ways.

For instance, many householders pay little heed to water supply pipes, blithely believing them capable of providing all the water a house could possibly need. In truth, any pipe has its limits of flow—a fact that may become sadly apparent to a homeowner only after he adds a shower, clothes washer or perhaps a new wing to his house, and discovers that the addition is starving for water.

Fixtures farther from the water source do not get as much water pressure as those that are nearer *(page 16B)*. Too-small pipes may aggravate this problem. In some older houses, the main supply pipe in the basement may be only ½ inch in diameter, rather than the ¾-inch size that should be the rule. If new fixtures are added to such a house, a larger main supply pipe may have to be installed from the water inlet to the various branches of the supply network in order to maintain satisfactory pressure.

If the inadequacy of a supply line may be overlooked for a time, so too can some of the problems with drainpipes. Of course, no one can miss for long their commonest failure—clogging—but they also are subject to some little-known maladies. Under certain circumstances, drainage can actually work in reverse: a process called back-siphonage *(page 16C)*, can suck waste water into a supply inlet, polluting the fresh-water supply. Such mishaps are fortunately rare, but, as with all aspects of a plumbing system, the surest preventive for trouble lies in understanding the pipes' unseen dynamics—latent or otherwise.

Four webs of pipes. In this simplified model of a plumbing system, fresh cold water enters the system via the basement *(blue pipe)* and is distributed to fixtures throughout the house. Some cold water is diverted to the water heater *(lower right)*, from which hot water flows *(red pipes)*. From each fixture, a down-sloping drain *(orange)* carries water to the soil stack *(vertical orange pipe)* and the main drain *(horizontal orange pipe)*. Vent pipes *(green)* keep air flowing through the drains to the atmosphere.

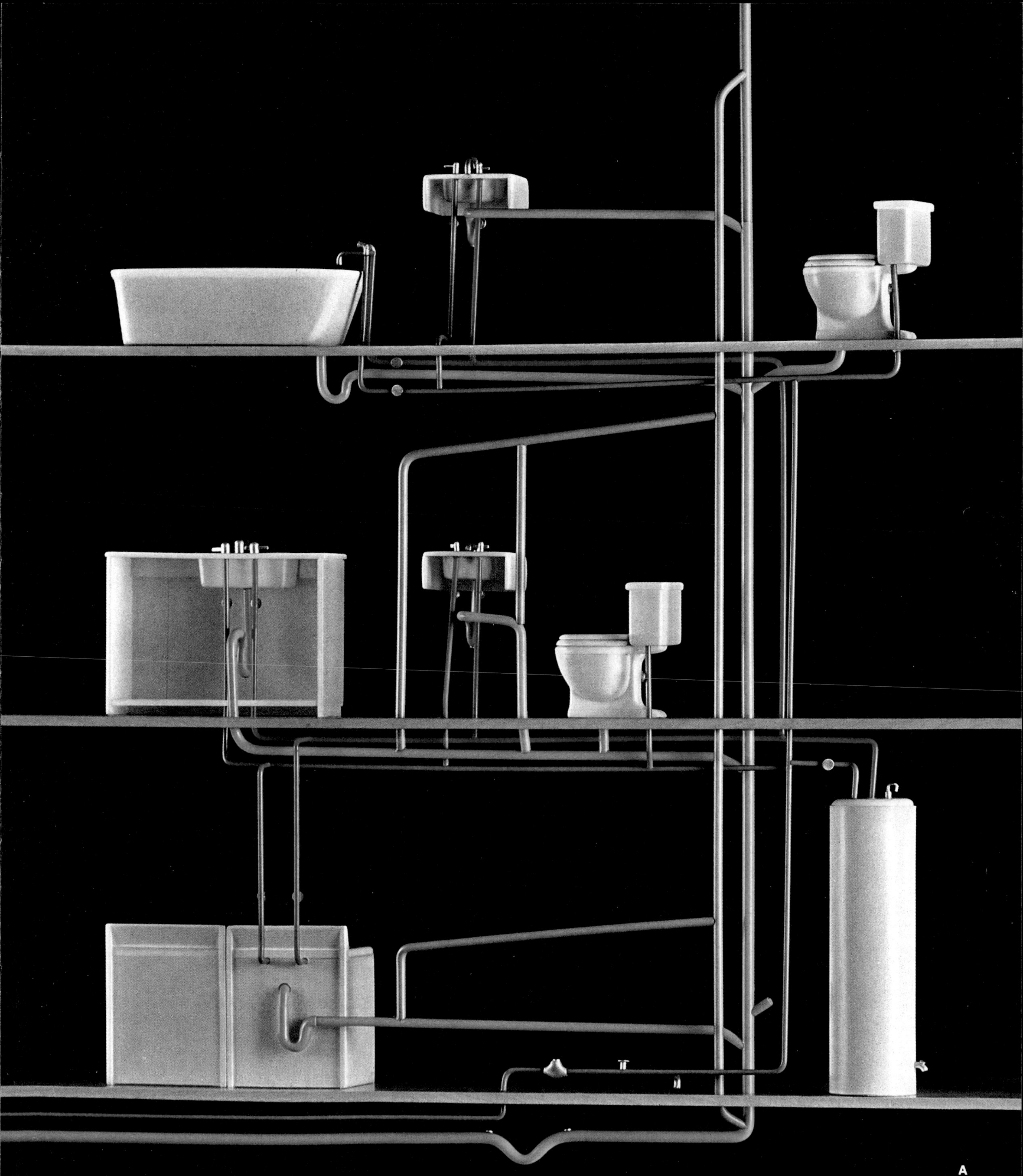

A

Far from being a reliable constant, water pressure varies greatly when water flows to different combinations of fixtures, through passages of different sizes, or if water freezes. The penalties of pressure-related problems, shown on these and the following pages, can range from inconvenience to plumbing disaster.

Pressure fall-off. In this experiment, the horizontal tube represents the main supply line, the tall vertical tubes stand for branch supply lines to fixtures and the short tubes represent the fixtures' outlets. When all the fixtures are turned off *(inset)*, pressure is everywhere equal. When all the fixtures are on *(large photograph)*, the pressure drops markedly the farther away a fixture is from the source of supply *(far left)*.

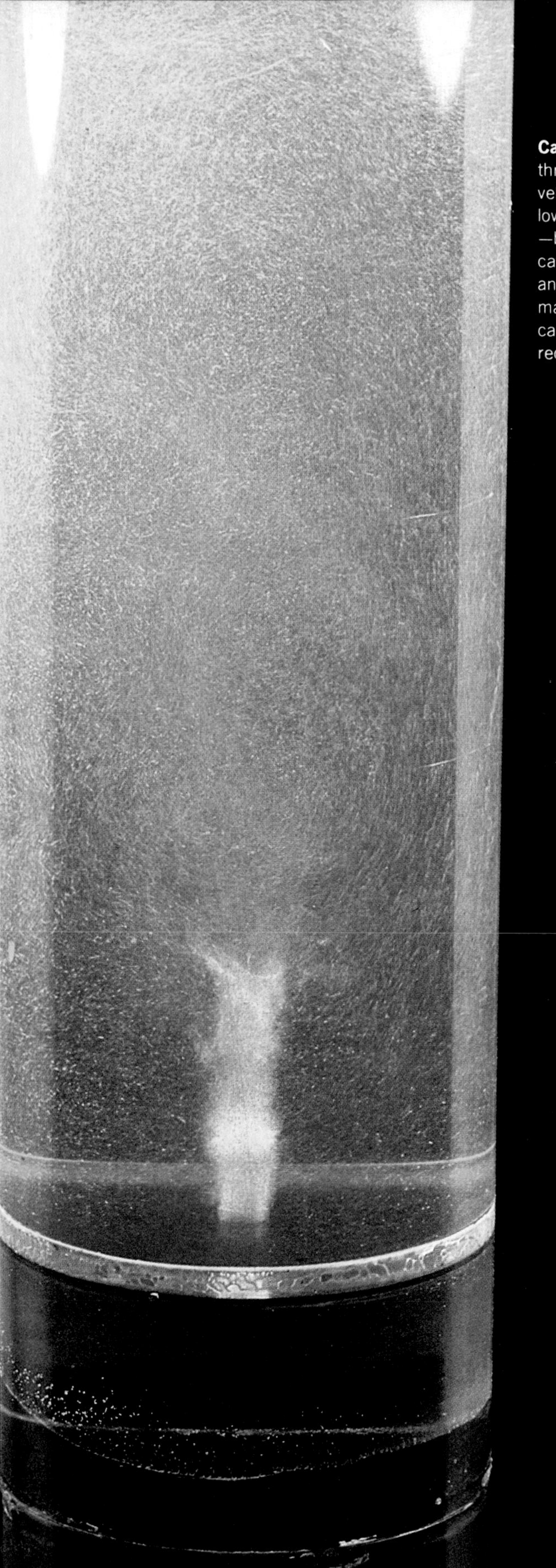

Cavitation. When water in a pipe passes through a small opening *(bottom of tube)* the velocity rises and the pressure drops. In the low-pressure area, gases—mainly oxygen—boil and form bubbles *(top)*, an effect called cavitation. Expanding bubbles generate an annoying screech, while collapsing bubbles may erode surfaces. Cavitation in a toilet valve causes noise and leakage; a pressure-reducing valve *(pages 102-103)* is the cure.

Water hammer. A multiple-exposure photograph shows how the banging of water hammer can occur if water flow is suddenly halted in a pipe with an elbow near some hard surface; in this demonstration the surface is a 2-by-4 stud, and the flow is controlled by a clothes washer's solenoid valve *(red fitting)*. When the valve is open, the pipe's elbow remains distant from the stud. But when the valve shuts off the flow, the water in the pipe continues to surge ahead for an instant, building up pressures that slam the elbow into the wood. The cure is a pressure absorber installed at the valve *(pages 102-103)*.

D

Destruction by ice. While most substances contract as they freeze, water expands—and bursts pipes. This is bad news for a homeowner who leaves water in exposed pipes in winter—or who suffers a furnace failure. Three stages in the freezing and rupture of a pipe are shown in the tubing. Frost forms on inner surfaces *(top)*, ice crystals take shape *(center)* and the pipe cracks. To prevent such disasters, drain outdoor piping *(page 17)* or use freezeproof valves *(page 36)* —and melt freeze-ups quickly *(pages 20-21)*.

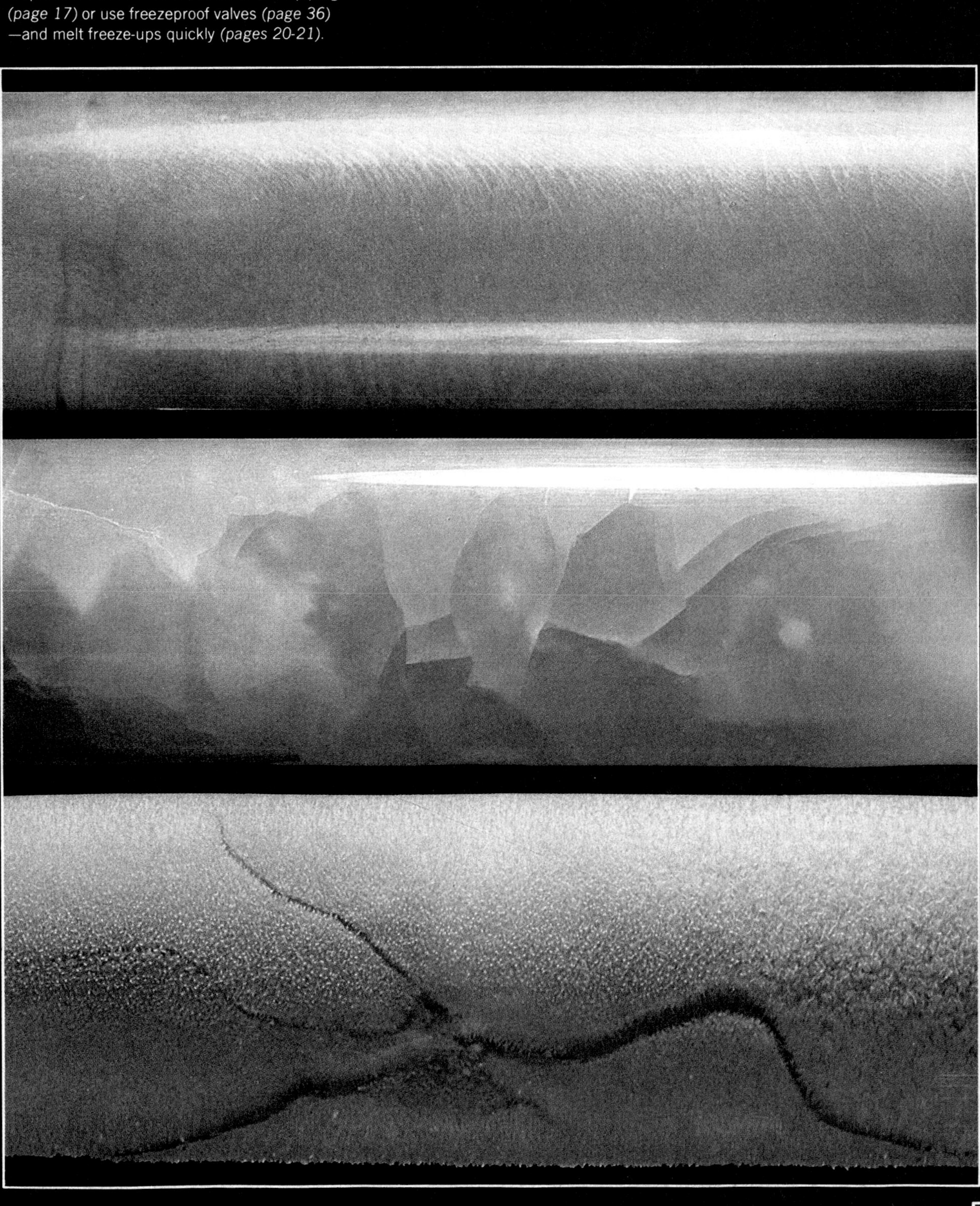

Drainage—Right and Wrong

Pulled by the force of gravity, used water leaves the home through extra-large pipes *(overleaf)*, while at each fixture a trap shuts the door on the departing waste *(below)*. Under some conditions, however, a drop in water pressure allows wastes to intrude into the supply system *(opposite)*, a menace called back-siphonage.

P-shaped protection. Under every lavatory, sink and bathtub, and inside every toilet, a bend in the drainage pipe—called a P trap because of its shape—stands guard against pollution. Water always remains in the bend of the trap, preventing gases *(dark gray)* from rising out of the drainage system and into the home. Many traps have a bottom plug for cleaning out obstructions—or for retrieving a ring that drops down the drain.

Back-siphonage. Waste water may acquire a means of access to fresh-water supply pipes if a turned-on spray shower is left submerged in a bathtub—represented here by a hose dangling in a container of black water. As long as pressure is maintained in the supply line, the waste water cannot enter *(inset).* But if there is a break in the supply main *(bottom, left),* the sudden drop in pressure may siphon the polluted water into the fresh-water side of the plumbing system.

Rings of water. Waste water flows down a vertical drain along its sides, clinging by the force of surface tension. If the water were to separate from the inner surface and fall freely—as extra-heavy flow can—noise and vibration would result. Proper annular, or ring-shaped, flow drags air down with the water, as indicated by the plume of smoke from the small hose in the center of this pipe. Vents supply such air from outdoors, so water is not sucked from fixture traps.

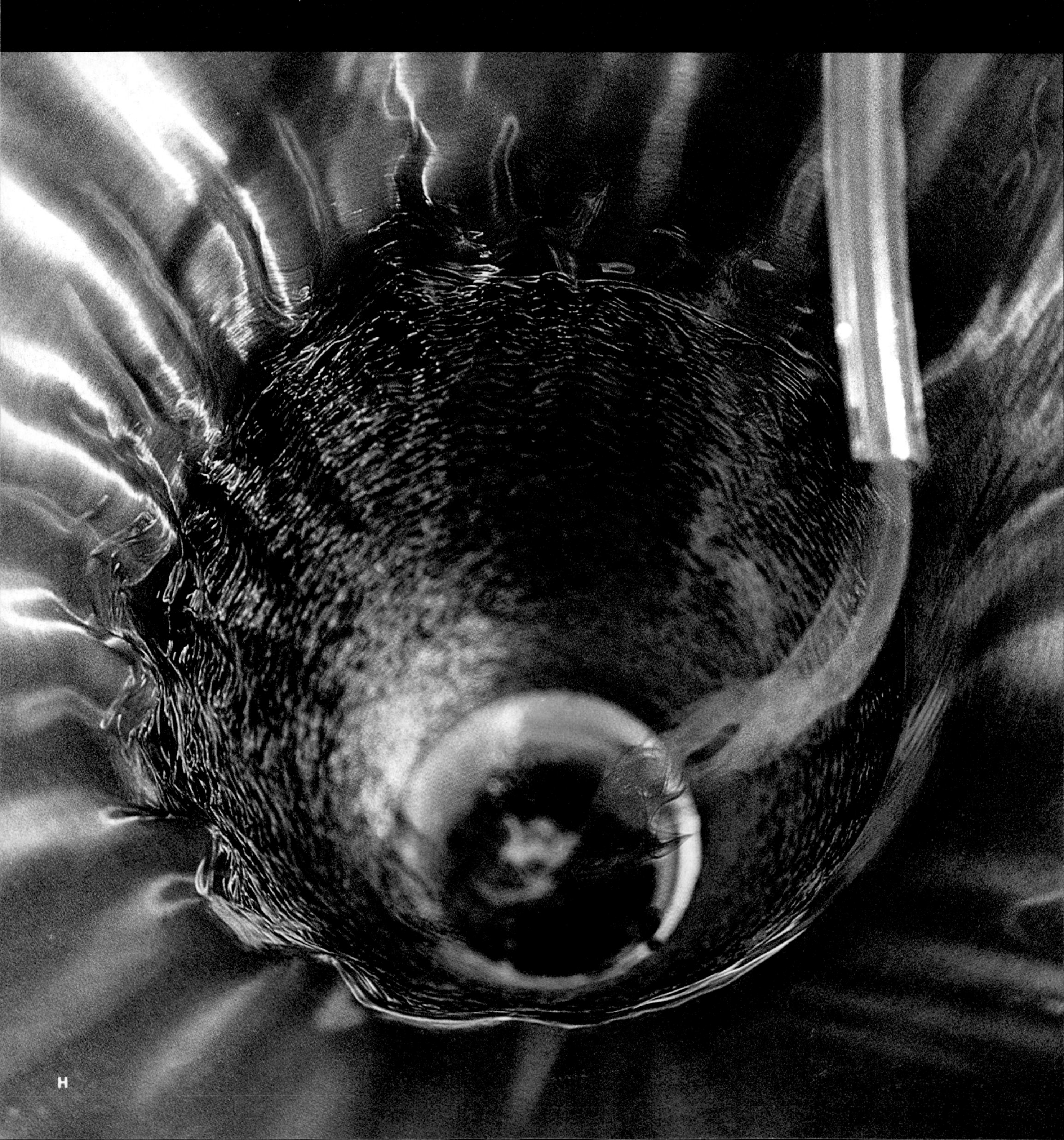

Shutting Down the Plumbing System

At one time or another, you may need to put all of the fixtures, appliances, faucets and pipes in your plumbing system out of operation. When you do major repairs or installations that call for cutting into the main pipes, or when you need to clear and recharge waterlogged air chambers *(page 102)*, you must both shut off the incoming water and drain out the existing supply before you can start work. When you leave a house empty and unheated for the winter, you must also weatherproof the system to protect it from bursting in freezing temperatures.

Whatever the occasion, be sure to reserve a few gallons of water for drinking or cooking before you shut off the supply. If you are about to winterize a summer home, draw enough extra water to prepare an antifreeze mix that you will need to protect the traps.

In order to drain your system efficiently follow this check list:

☐ Cut off the house water supply by closing the main shutoff valve.

☐ Turn off the gas or electricity to the boiler and the water heater.

☐ Siphon the water out of the tub of the clothes washer.

☐ If you have hot-water heat, open the drain faucet on the boiler and let the water flow into the floor drain. Next, open all of the radiator valves. Then remove an air vent from a radiator on the top floor so that air will replace the water as it drains into the boiler.

☐ Working floor by floor, starting at the top, open all hot and cold water faucets —including all tubs, showers and outdoor faucets—and flush all toilets.

☐ Open the drain faucets on the water heater and the water treatment equipment, if you have any.

☐ Finally, open the drain faucet on the main supply line to release any water that may remain in the pipes.

At this stage, your plumbing system will be adequately drained for repair or remodeling work. If you are closing the house for the winter, take additional precautions. Walk through the house to make sure every place where water can collect is drained.

For cold weather protection, the water still remaining in fixture and toilet traps as well as the main house trap, if you have one, must be replaced with an antifreeze solution to keep the traps from bursting while still functioning as a barrier against sewer gases. Get the nontoxic propylene glycol antifreeze sold for recreational vehicles; the ethylene glycol antifreeze used in automobiles is toxic (the manufacturer recommends that it not be brought into the house) and alcohol-based products evaporate too fast.

Mix the antifreeze with water, as directed on the label, in the same proportions you would use to protect a vehicle in your climate. (If you do not want to mix it, pour it straight from the can into the trap.) How much antifreeze you need will depend on the proportions that are recommended and the number of traps to be winterized.

Prepare the lavatory, sink and tub traps first. Remove all the accessible cleanout plugs, drain the water from each trap into a pail or bucket and discard it; replace the plug. Pour at least a quart of the antifreeze solution into each trap. With traps you were unable to empty, pour the solution in very slowly so that it will push the existing water ahead of it into the drainpipes.

Next, wipe up any water remaining in the bottom of toilet bowls with rags or newspapers. Pour at least a gallon of the antifreeze mixture into each toilet tank, then flush the tank to dislodge water from flushing channels of the toilet bowl. The antifreeze will collect in the toilet trap.

To complete the winterizing, remove either the inlet or outlet plug of the main house trap, if you have one, and siphon out the water in the trap. Pour about a quart of antifreeze into the trap and replace the plug. If you have your own well system, drain the water tank and dry off all parts of your pump—unless it is a submerged one, which requires no special precautions.

Come spring, or after the repair is completed, you will have to refill your system. First, close the drain faucet on the main water supply line. Close all the faucets throughout the house including those on the boiler, hot-water heater and water treatment equipment. If you have hot-water heat, replace the air vent on the radiator from which it was removed. Then open the main valve slowly to bring fresh water into the system. Finally, turn on the gas or electricity supply to the boiler and hot-water heater and light the pilots on gas-fired equipment.

Faucets will sputter when you first use them because of air trapped in the pipes, but this condition will correct itself quickly. The antifreeze mixture in the traps and toilets will flush away naturally as the fixtures are used.

Emergency Pipe Repairs

Pipes have a disconcerting way of springing a leak on Saturday night, after your friendly plumber has left for the annual union banquet. Fortunately, some simple fixes, easy to cobble together from the junk box on a workbench, will tide you over until you can make (or commission) a permanent repair *(pages 70-81)*.

When you discover a leak in a supply pipe, the first thing to do, of course, is shut off the water. This will reduce pressure on the damaged section so you can proceed to plug the hole.

An application of epoxy glue or plastic tape is the quickest emergency procedure. Various "bandages" or plugs are even better. But before trying any remedy, make sure the pipe surface is dry enough for adhesives or sleeves to hold.

In the case of a leaking supply pipe that is not frozen, completely drain and dry the affected section if possible—an electric hair drier does a quick job. Damaged pipes that are frozen should be left unthawed and undrained until patching is completed *(pages 20-21)*. Drainpipes and traps, unlike supply pipes, are not under pressure and normally contain no water unless a fixture is in use.

A Patch for a Tiny Leak

1 Plugging a hole with a pencil. One of the best emergency plugs for a small leak in a supply pipe is a pencil point jammed into the hole and broken off; the soft graphite point will conform to the shape of the opening and seal the leak.

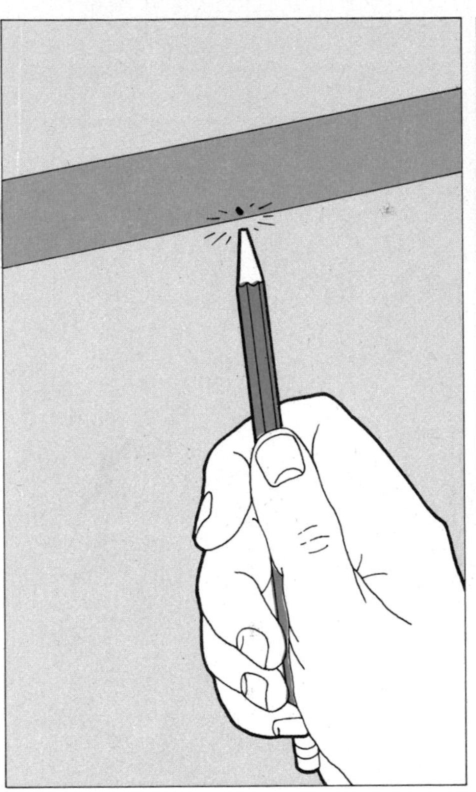

2 Securing the plug. Dry the surface of the pipe after the leak has been plugged, then roll heavy tape over the damaged area to hold the plug in place. Wrap the tape several inches to the left and the right of the leak.

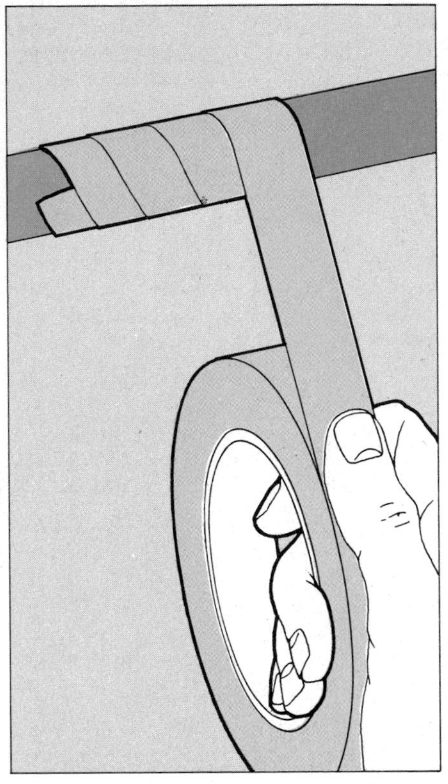

A Patch for a Larger Leak

A hose patch. An effective temporary patch can be made by splitting a section of rubber hose lengthwise so that it will fit around the pipe. There are several ways to secure the hose. Strong, flexible wire such as that used for hanging pictures will serve—make a series of loops along the patch, spaced about an inch apart, and twist each loop tight with pliers *(right)*. An automobile hose clamp *(far right, top)* holds more uniformly and can be adjusted to fit virtually any diameter of fresh-water pipe; it is best to install at least three clamps over the patch. To guarantee uniform clamping pressure, cut a section out from a tin can wide enough to cover the leak and long enough so the top rims of the sheet metal extend above the pipe when fitted around the hose patch. Bend the top rims at a right angle to the pipe so they fit closely together. Drill holes through both rims and fasten with bolts and nuts.

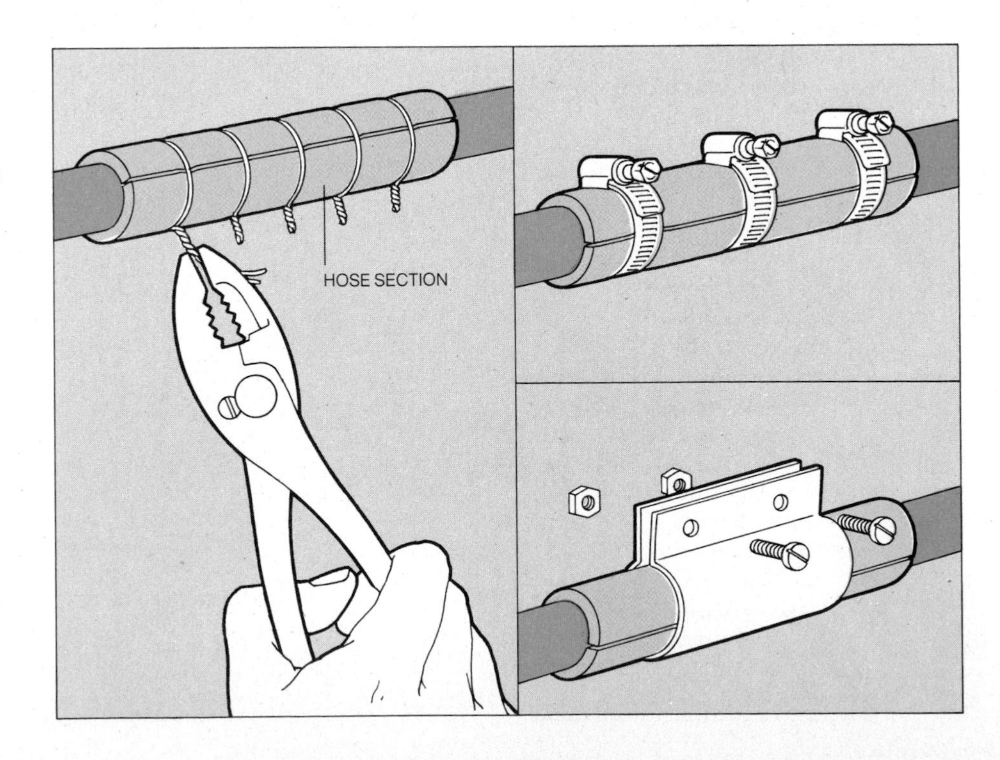

A Semipermanent Patch

Installing a pipe sleeve. If you postpone replacement of a damaged pipe for any length of time, replace a homemade clamp as soon as possible with a pipe sleeve, available in plumbing-supply stores. Simply fit the sleeve around the damaged section of pipe and tighten the nuts and bolts. Such sleeves combine patch and clamp in one unit and come in various sizes; get one that fits the diameter of the pipe you are patching.

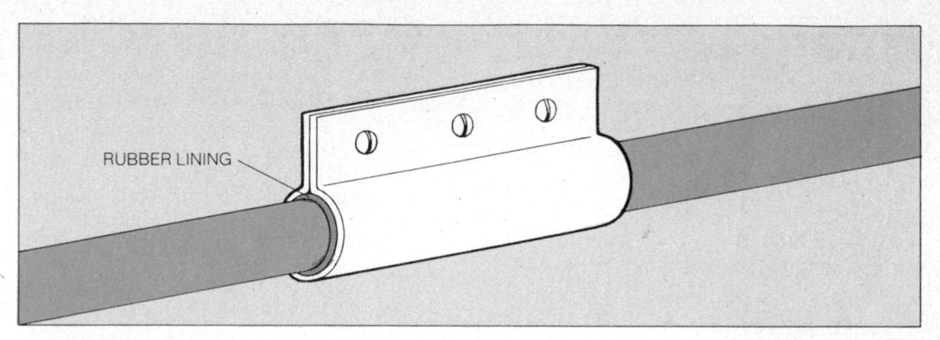

Minimizing Water Damage if a Leak Floods You Out

Often the first sign of a leaking pipe will be a spreading stain on a wall or ceiling or a puddle on the floor. To prevent further damage, shut off the water supply immediately, before trying to trace the leak and repair it.

You can sometimes anticipate water damage and keep it to a minimum. Where leaking pipes are concealed above the ceiling and a water stain is visible, place a waterproof dropcloth on the floor and position a catch basin under the wet area. If water is leaking from a ceiling light fixture, shut off the electricity and drain the fixture by removing its cover. Poke a hole through the ceiling or remove a section of it to let any remaining water drain out. Stand out of the way!

During a plumbing freeze-up, take precautions against leaks until you can be certain the pipes have suffered no damage. Since the leaks will be frozen until the pipes thaw, waterproof the suspect area with plastic dropcloths like those used by painters. If you spot a crack, put a clamp on it *(pages 18-19)*. And be ready with a few extra pots and pails in case undetected leaks reveal themselves.

If you arrive on the scene too late to avert a flood, you can still construct a makeshift dam from sandbags or rolled-up rugs to prevent the flood from spreading to other rooms. For a bad flood, you may need a pump with a submersible motor, usually available from rental agencies (check under "pumps" in your classified phone directory). If the situation is desperate, call the fire department—which is usually prepared to help.

Leaks at Joints

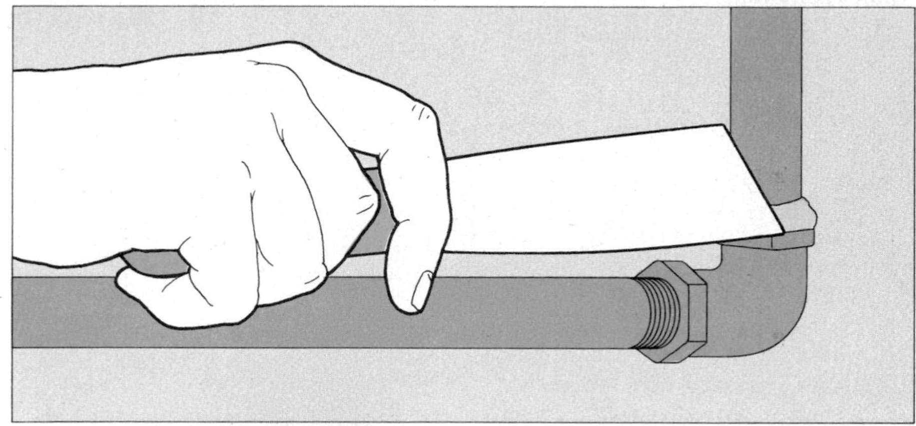

Binding threaded pipes. If a leak appears at a threaded joint, drain the pipe, dry the damaged area and apply epoxy cement over the leaking joint. Allow the epoxy to harden completely, according to the manufacturer's instructions, before restoring water pressure. Epoxy also can be used on cemented plastic joints.

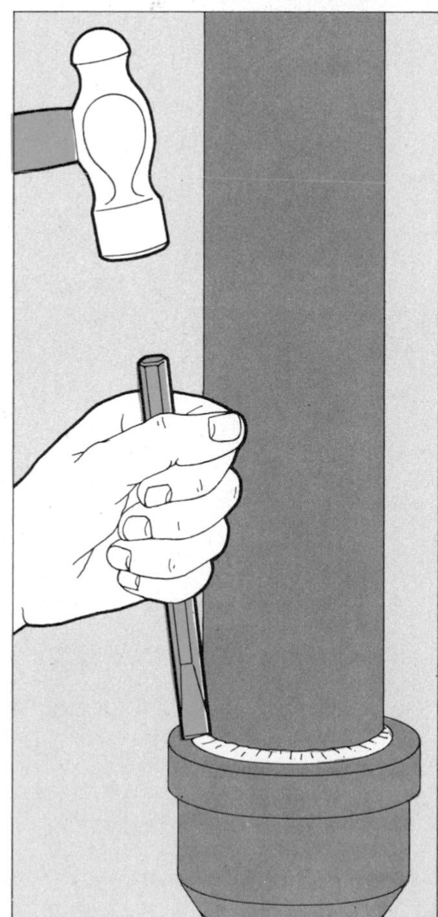

Fixing a lead-caulked drain joint. If water seeps from a lead-caulked drain joint, use a hammer and chisel to tamp down the lead inside the hub of the drain pipe. Since the lead is soft enough to be reshaped over the weak spot, this simple procedure often is sufficient to reseal the joint.

First Aid for Frozen Pipes

A properly constructed and heated house is safe from plumbing freeze-ups even during the most severe cold snap—unless the heating system breaks down or is knocked out by a power failure. Then no home is immune. If heat is not restored promptly, indoor temperatures will drop precipitately and you must act quickly to keep pipes from freezing and bursting *(right)*. Even in an otherwise well-built house, pipes that run through an unprotected crawl space, basement, laundry room or garage can freeze during exceptionally cold weather, especially if the room is drafty.

If pipes do freeze, the first sign may be a faucet that refuses to yield water. But all too often, the freeze-up is announced by a flood from a break. Water expands about 8 per cent in volume as it begins to freeze, generating pressure that splits pipes, especially where expansion is impeded by joints or bends. Ice may form throughout a long straight section of supply line before it meets an obstruction and cracks the pipe; thus, the entire length of pipe that supplies a stopped faucet should be considered suspect, both for ice blockages and leaks.

When you prepare to thaw a section of pipe, keep the affected faucet open to let vapor and melting ice run out. Then turn off the water supply once you have located the leaks and marked them for repairs *(pages 18-19)*. After temporary patches have been applied to the damaged areas, open the main shutoff partway; the movement of water through the frozen section of pipe will aid the thawing process. The surrounding area should be guarded against water damage *(box, page 19)* in the event that any other leaks have gone undetected.

Electrical heaters of one kind or another are generally safest for thawing. However, freeze-ups often occur during power failures and in such a case you are likely to have to use the flame from a propane torch.

Electric heating tape. Wrap the tape in a spiral around the frozen pipe, allowing about six turns per foot, and leaving at least half an inch between each turn. Secure the spiral with masking tape about every 4 inches. Some electric heating tapes come with built-in thermostats and can be left permanently plugged in: when the temperature drops toward the freezing point, the thermostat automatically activates the tape and warms the pipe.

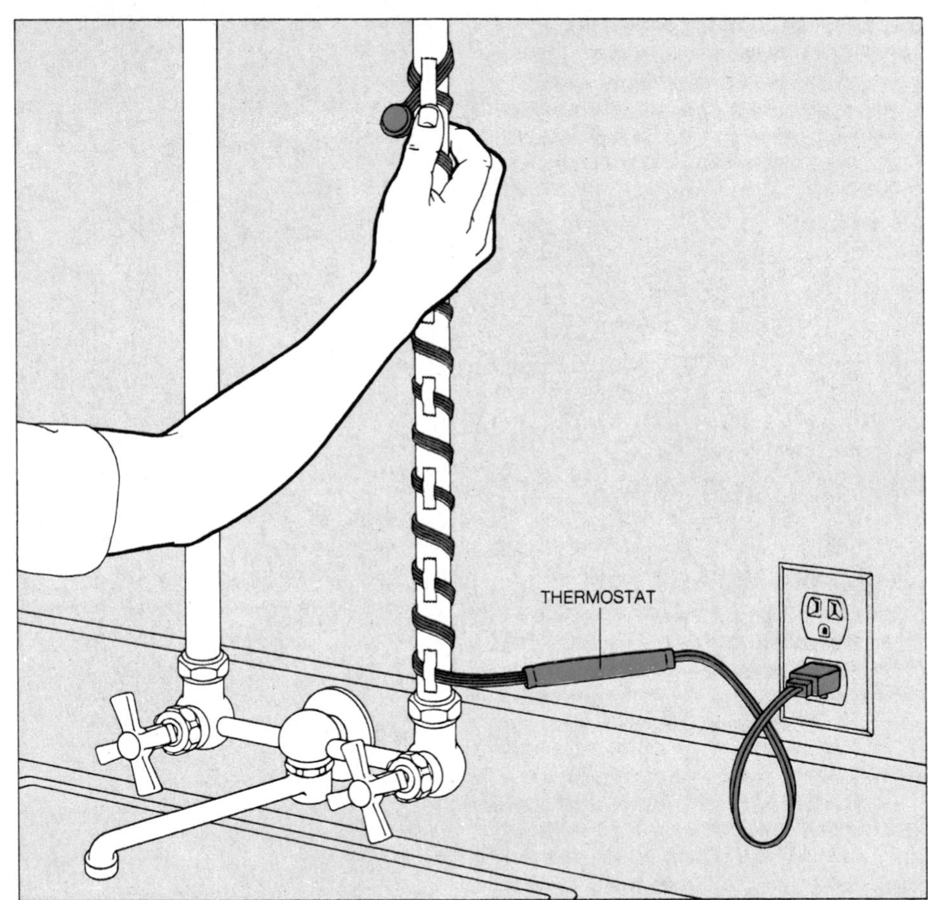

How to Keep Pipes from Freezing

There is only one sound way to prevent freeze-ups in an unheated house: drain the entire plumbing system as described on page 17.

For pipe protection in other circumstances, you can choose from several methods—both electrical and nonelectrical, temporary and permanent.

☐ If power is available, plug in an electric heater or heat lamp, or hang a 100-watt bulb near vulnerable pipes.

☐ Keep a door ajar between a heated room and an unheated room with pipes so that the unprotected area will receive heat.

☐ Set an electric fan on your furnace to blow warm air over basement pipes.

☐ Insulate exposed pipes. In addition to thermostatic heating tape, there are wrap-on and snap-on types of insulation that protect vulnerable pipes.

☐ If no commercial insulation is at hand and pipes must be protected immediately, wrap several layers of newspaper loosely around the pipes and tie with string.

☐ If the temperature suddenly drops and you have no time to install insulation, turn faucets on to a trickle; this will retard freezing.

Propane torch. Equipped with a flame-spreader attachment, a propane torch can thaw a pipe rapidly and effectively, but use it cautiously. Place asbestos sheeting between the pipe and nearby framing. Apply heat near the open faucet first, and then work gradually back along the pipe, feeling frequently to make sure that you are not overheating it. The pipe should never become too hot to touch; if you get it so hot that water boils inside, steam can cause a dangerous explosion.

Hair drier. If you have electricity, an appliance that blows warm air—a hair drier or a tank-type vacuum cleaner with the hose set into the outlet end—can be used in the same fashion as a propane torch, although it will work more slowly.

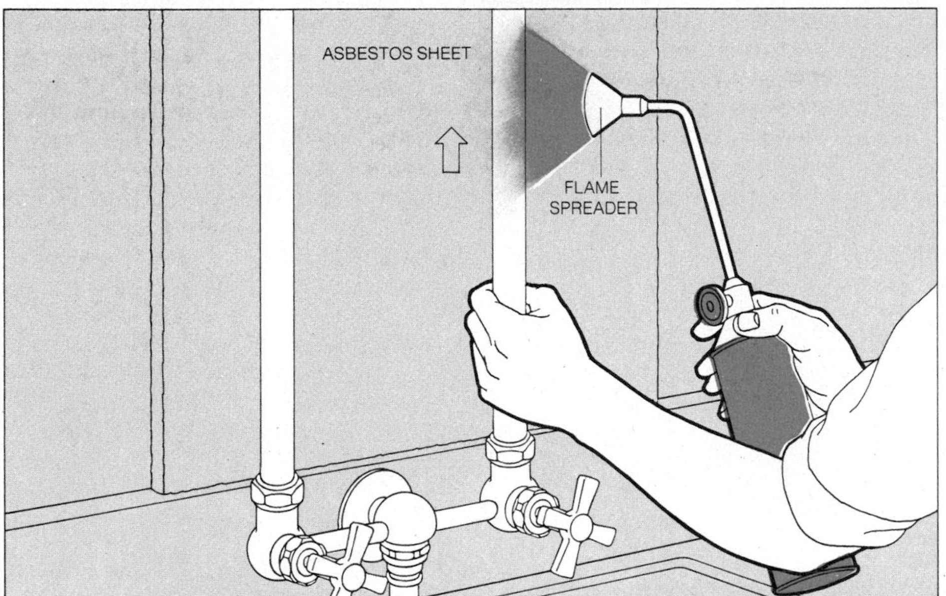

Heating pad. Wrapped and tied around the frozen pipe near the faucet, an ordinary heating pad thaws ice slowly but effectively. Leave it in place until water stops dripping from the faucet, then move it to thaw another section.

Hot water. Tie rags or towels around the frozen section of pipe and soak them with water from a kettle. Continue pouring a light stream of water over the rags until the ice has melted.

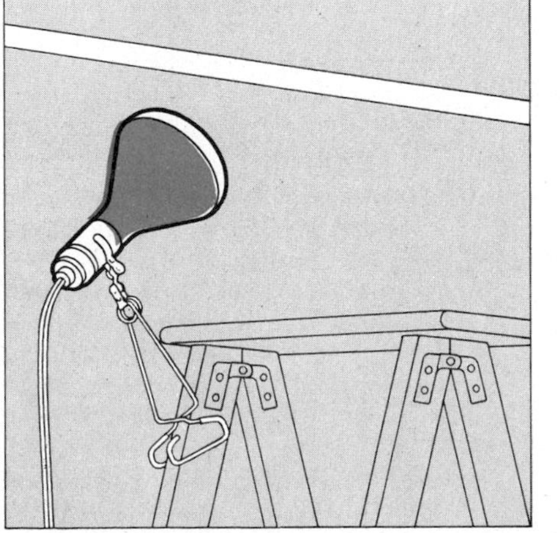

Heat lamp. If the suspected ice blockage is behind a wall or otherwise out of reach, set an electric heat lamp nearby. Keep it at least 6 inches from the wall to avoid scorching paint or wallpaper. For greater flexibility in handling, you can screw the bulb into the socket of a portable work lamp. Pipes that disappear into the ceiling can also be thawed with a heat lamp; hang the lamp directly below the frozen section.

Time-tested Methods for Unclogging Drains

A drain blockage may be in the trap under the fixture, in a branch line carrying waste from several fixtures to the main drain, or in the main house drain or its connection to the sewer *(page 16A)*. If only one fixture is blocked or sluggish, start by cleaning its trap. If several fixtures are clogged, follow the techniques on the opposite page and pages 26-27 for cleaning branch and main drains. When working with branch or main drains, turn off the water at the main shutoff valve to avoid flooding if someone should turn on a faucet elsewhere in the house.

A common force-cup plunger and a drain-and-trap auger are not only the easiest unclogging tools to use, but also the most effective and the safest. Devices that use compressed air often impact the blockage and may loosen or blow apart fragile pipes such as lavatory traps.

It is not advisable to use a chemical drain cleaner in a fixture that is completely blocked. The most powerful cleaners contain lye, a caustic compound that is dangerous and can harm fixtures if left too long in them. If the chemical cleaner does not clear the drain, you will be exposed to the caustic when you open a cleanout plug or remove a trap.

After a drain has been cleaned, chemical cleaners do serve a useful purpose. When used regularly—every two weeks or so—they prevent build-up of debris that could lead to a future blockage.

Lavatories and Sinks

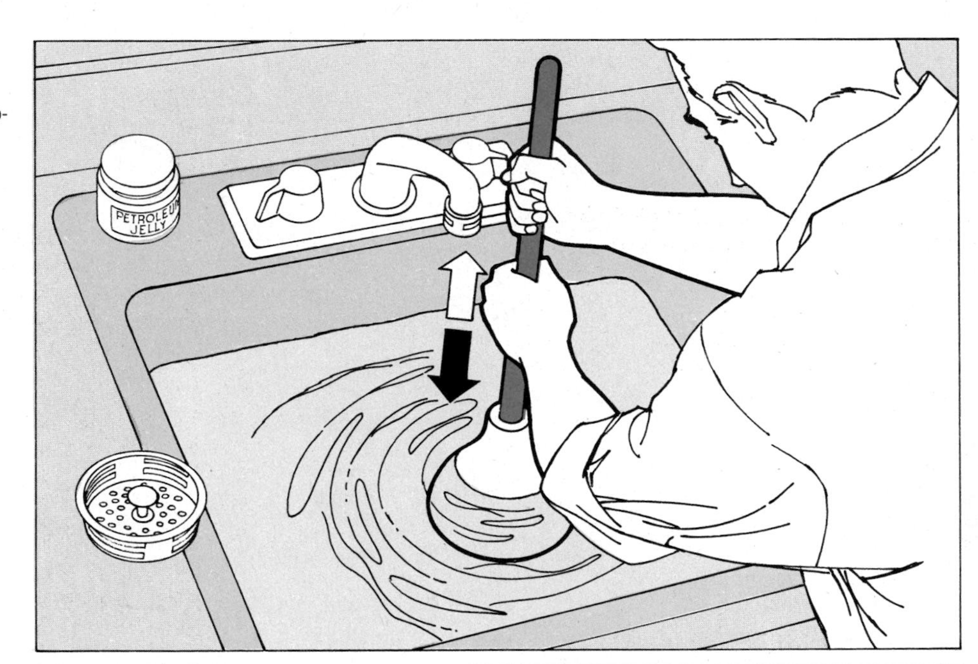

1 Using a plunger. Remove the sink strainer or pop-up drain plug *(pages 48-49)*. If there is an overflow opening—most kitchen sinks have none—plug it with wet rags. Be sure there is enough water in the basin to cover the plunger cup completely. Coat the rim of the cup evenly with petroleum jelly and center the cup over the drain hole. Without lifting the cup, pump down and up with short, rapid strokes 10 times, then jerk the plunger up from the drain quickly. Repeat the procedure several more times if necessary. If the drain still remains clogged, try Step 2.

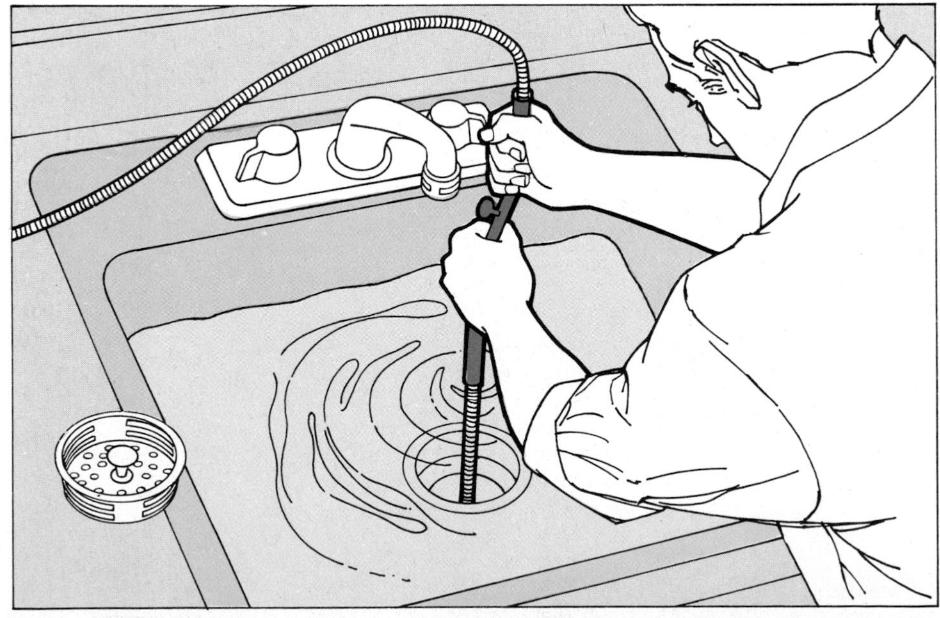

2 Using an auger. With strainer or pop-up stopper removed, feed a drain-and-trap auger into the drain by cranking the handle clockwise. As you push the auger wire farther into the drain, alternately loosen and tighten the thumbscrew on the auger handle. When you hook something move the auger backward and forward slowly while cranking, then withdraw the auger wire slowly while continuing to crank in the same direction. Pour hot water and detergent through the trap to clear away residual grease or oils. If the auger does not clear the trap, try Step 3.

3 Working through the cleanout plug. If the trap under the clogged fixture has a cleanout plug, place a bucket under the trap and remove the plug. After water has emptied from the trap, straighten a wire coat hanger, form a small hook in one end, and probe through the trap. If the obstruction is near the opening, you should be able to dislodge it or hook it and draw it out. If not, feed a drain-and-trap auger first up to the sink opening, then through the back half of the trap. If the blockage is not in the trap, try Step 4.

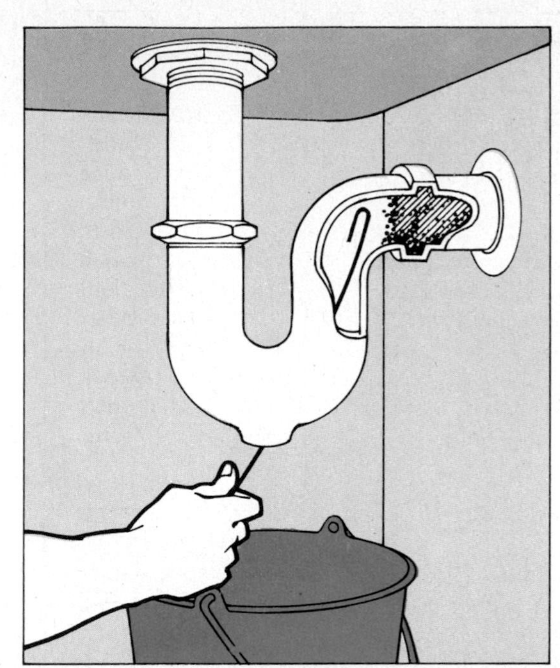

4 Removing a trap for cleaning. A trap may not have a cleanout plug. If the blockage cannot be cleared by a plunger or an auger inserted through the sink drain, shut off the water, wrap the jaws of a wrench with tape to protect the chromed slip nuts at either end of the trap, unscrew the nuts and detach the trap. Clean it with detergent and a bottle brush. Replace the trap and turn on the water. If the drain is still clogged, try Step 5.

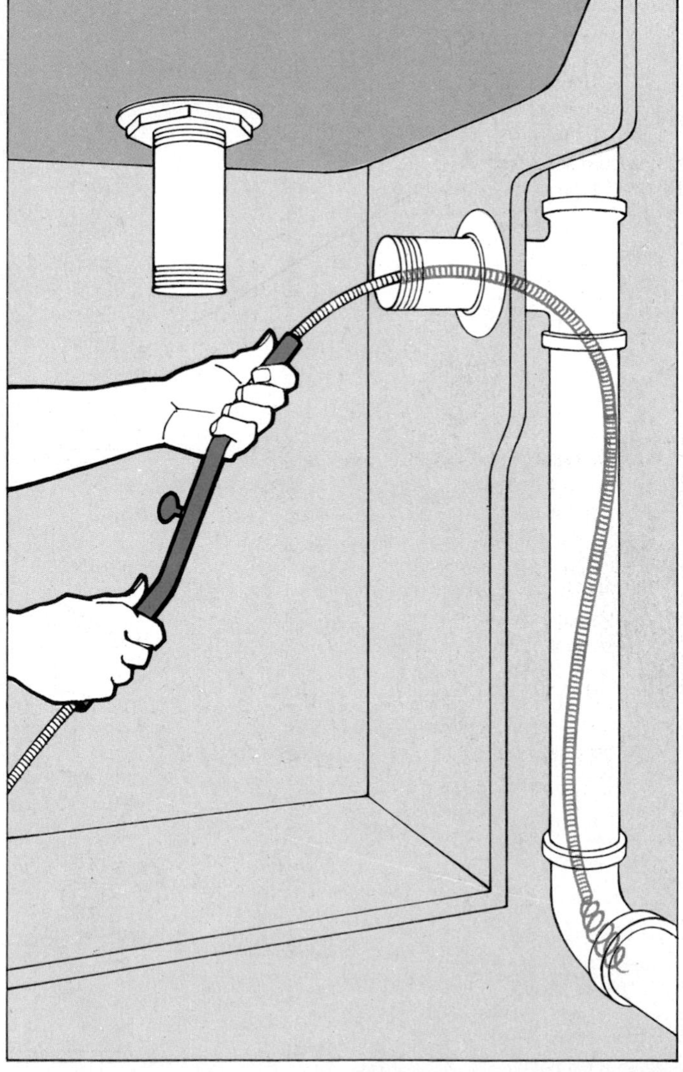

5 Cleaning beyond the trap. With the trap removed, crank the drain-and-trap auger into the exposed end of the drain pipe that goes into the wall. The blockage may be in the vertical pipe behind the fixture or in a horizontal pipe—a branch drain—that runs through the wall or floor to connect to the main drain-vent stack serving the entire house. If the auger goes in freely through the branch drain until it hits the main stack, the blockage is probably in a section of the main drainage system *(pages 26-27)*.

Drains below Floors

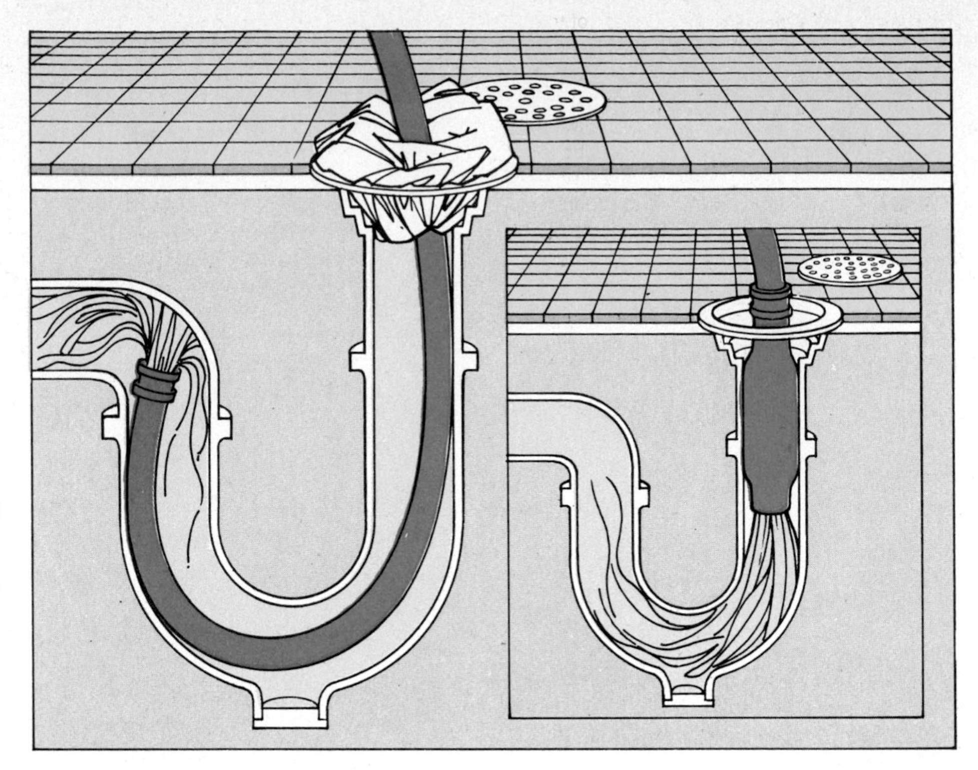

A hose instead of an auger. For floor drains like those in showers or basements, a garden hose often is more effective than an auger if the blockage is far down. Remove the strainer from the drain. Attach the hose to a faucet (a threaded adapter is needed for a sink or lavatory faucet). Push the hose into the drain and pack rags around it. While you hold the hose—and the rags—in the drain, have a helper turn the water alternately on full force and abruptly off. The surges of pressure should clear the blockage.

To get increased pressure you can buy, at plumbing-supply stores, an inexpensive rubber device *(inset)* that resembles a hose nozzle and seals the hose in the drain better than rags. When the water is turned on, the device expands against the drainpipe so that the full force of water is directed into the drain. Caution: Never leave a hose in any drain. If water pressure should drop suddenly and drastically—a rare but not impossible occurrence *(page* 16G)—sewage could be drawn back into the fresh-water system.

Bathtubs

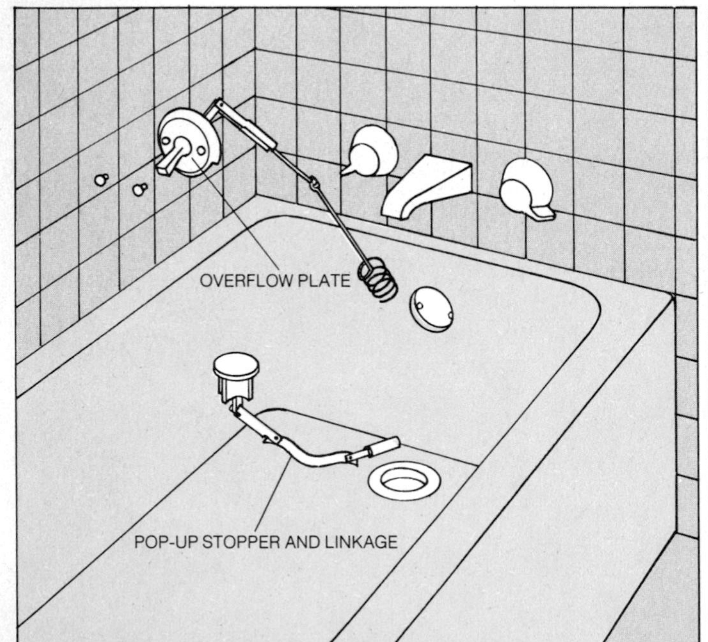

1 **Gaining access to the drain.** To unclog a bathtub, unscrew the overflow plate and lift it up and out. Draw out the stopper and its linkage. (Note how the parts line up so that you can put them back in the same way—there is a detailed view of the linkage on page 53.) Close the overflow opening with rags or masking tape, then use a plunger to open the clog. If that fails, try an auger as explained in Step 2.

2 **Unclogging a bathtub with an auger.** Run the auger through the tub overflow opening to reach the P trap, which serves both tub drain and overflow. Use the cranking and back-and-forth movements of the auger described on page 22.

A different type of bathtub trap. Instead of the common P trap, bathtubs in older houses and apartments may have a so-called drum trap located at floor level alongside the tub. To get at it, unscrew the cover of the trap counterclockwise with an adjustable wrench. Remove the rubber gasket. Using a drain-and-trap auger, first search for a blockage in the lower pipe inside the drum. If you find no obstruction there, insert the auger in the upper pipe, which goes to the main drain.

Toilets

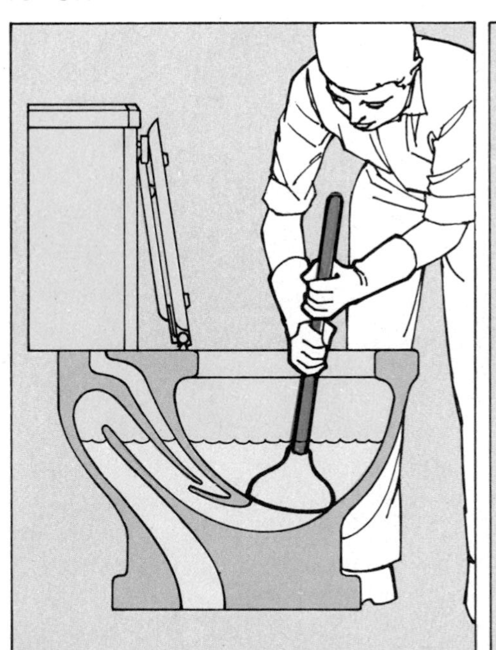

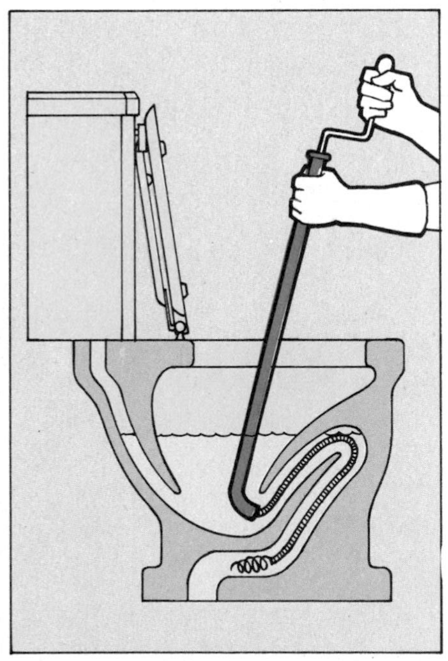

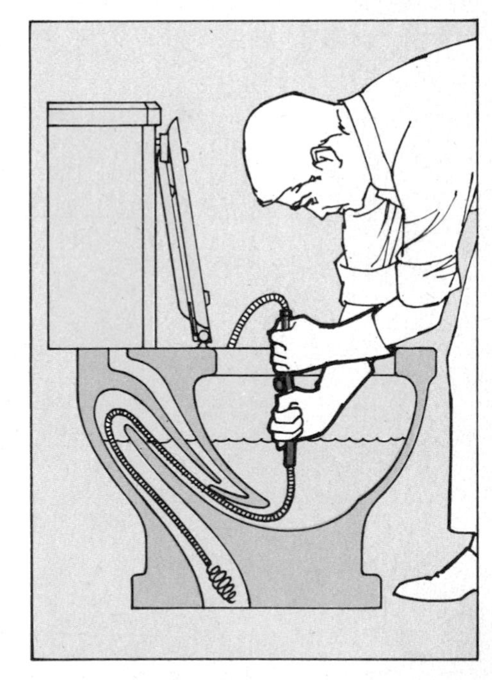

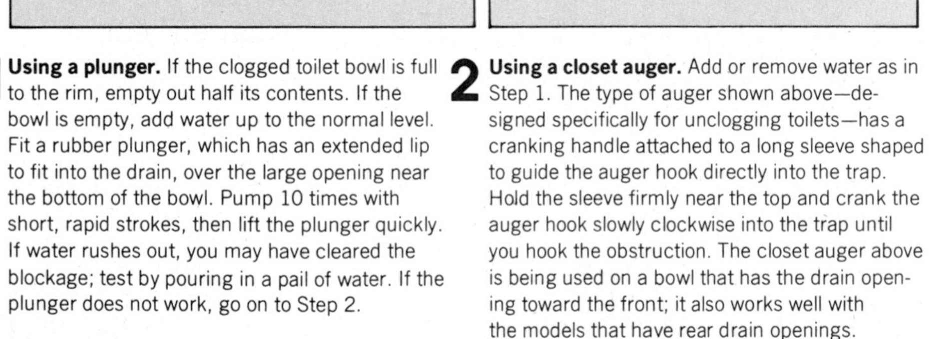

1 Using a plunger. If the clogged toilet bowl is full to the rim, empty out half its contents. If the bowl is empty, add water up to the normal level. Fit a rubber plunger, which has an extended lip to fit into the drain, over the large opening near the bottom of the bowl. Pump 10 times with short, rapid strokes, then lift the plunger quickly. If water rushes out, you may have cleared the blockage; test by pouring in a pail of water. If the plunger does not work, go on to Step 2.

2 Using a closet auger. Add or remove water as in Step 1. The type of auger shown above—designed specifically for unclogging toilets—has a cranking handle attached to a long sleeve shaped to guide the auger hook directly into the trap. Hold the sleeve firmly near the top and crank the auger hook slowly clockwise into the trap until you hook the obstruction. The closet auger above is being used on a bowl that has the drain opening toward the front; it also works well with the models that have rear drain openings.

Using a trap-and-drain auger. Insert the coiled-spring end into the drain opening. To avoid chipping the fragile vitreous china of the bowl, you must guide this auger carefully past projections of the trap. Crank the auger handle clockwise until you break through the obstacle or hook onto it and can pull it out. When the drain seems clear, test with a pail of water.

Unclogging the Main Drains

If all the waste from the fixtures in a bathroom backs up into the tub, or water from the clothes washer floods the basement, the trouble probably lies in the main drain and its branches, which channel waste into the sewer. When a clog forms in any part of the major drainage system, all the fixtures above the blockage stop up. When a clog forms in the stack vent, which keeps air flowing through the system, waste drains sluggishly and odors may be noticeable in the house.

Before you start work, track the pipes and connections in your own, perhaps unique, drainage system and try to pinpoint the trouble. If the clog seems to be in a branch drain, you can clean it out with a drain-and-trap auger, working at the fixture closest to the soil stack and following the instructions on page 23, Step 5. In order to make the job easier, you can rent an electrically powered auger up to 50 feet long from a plumbing-supply dealer.

Even an electric auger, however, will not work effectively if you push it around too many bends in the pipes. So if the problem seems to be beyond a branch, take the straightest possible route to the clog. To clear the main stack, the vertical pipe to which branches are attached, feed the auger down from the vent on the roof; to clear the basement or below-floor-level drain connected to the bottom of the stack, work through the main cleanout or—if you have one—the house trap. When you suspect that the problem is in the sewer line, somewhere between the house and the street, you will most likely need the services of a professional plumber.

In most plumbing repairs speed is essential, but unclogging the main drain calls for patience. The column of water trapped by the blockage may extend above the cleanout, and it will gush out, adding to the existing flooding, as soon as the plug is removed. If possible, wait for at least two or three hours after you spot the trouble to permit dispersion of whatever waste can seep past the blockage. Even then, you will need to arm yourself with mops, pails, rags and old newspapers for soaking up the overflow.

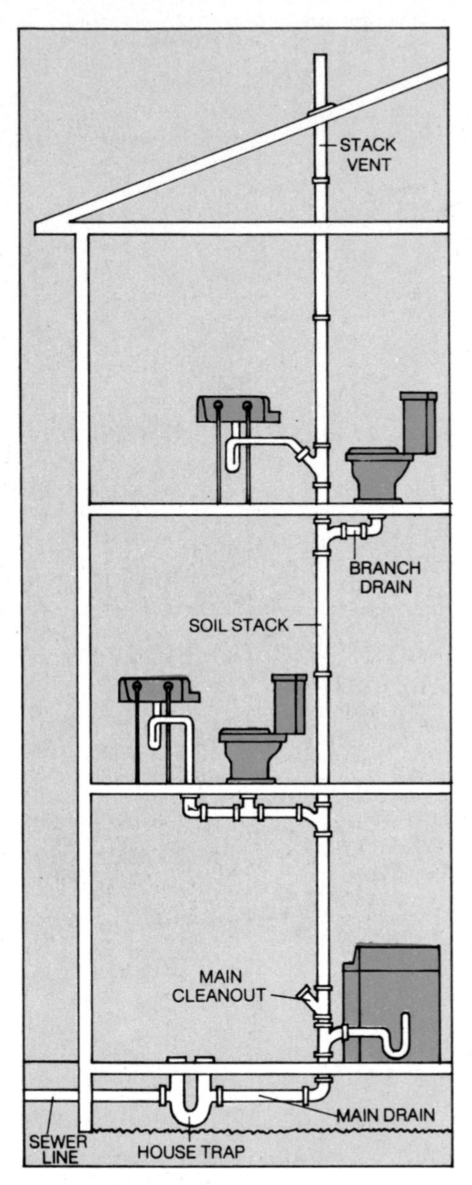

Locating the clog. Since waste flows downward and toward the sewer, any clog in a major drain is always below the level of the lowest stopped-up fixture and above the level of the highest working fixture. In this example, if the fixture at top left drains freely while the one at top right is blocked, the blockage is in the branch drain at right. If all top floor fixtures clog and the main floor and basement fixtures drain freely, the blockage is in the upper soil stack. If the main floor and top floor fixtures are stopped up while the basement ones drain freely, the blockage is in the lower soil stack. Clear the stack from the roof *(right)*. If everything clogs, the blockage is in the main drain; clear it from the main cleanout or house trap *(opposite)*.

Clearing the Soil Stack and Stack Vent

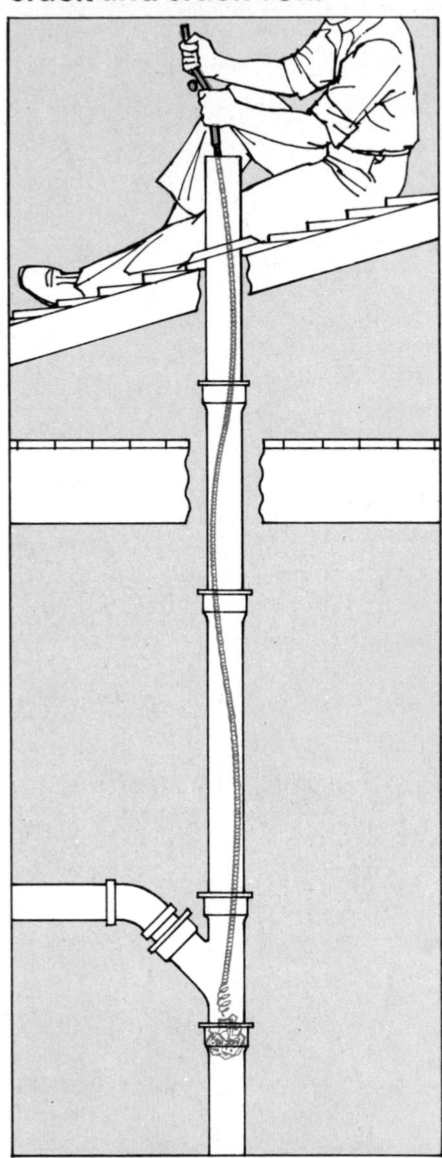

Running an auger down the stacks. Choose a drain-and-trap auger that is long enough to reach to the bottom of the soil stack. Position yourself securely on the roof beside the vent opening; use extreme caution when working on a roof. Feed the auger down the stack, using it as described on page 23; then flush out the stack with a garden hose.

Getting at the Main Drain from the Cleanout

1 Opening the main cleanout. Look for this Y-shaped fitting near the bottom of the soil stack or where the drain leaves the house. Set a pail underneath the cleanout plug and lay rags around to catch the flood that may occur. Use a pipe wrench to unscrew the plug counter-clockwise. If the plug does not turn, first try working penetrating oil into the threads, then slide a section of pipe over the wrench to increase your leverage. As a last resort, nick the plug's edge with a cold chisel and—keeping the blade in the nick—hammer the plug around.

2 Working from the main cleanout. Remove the plug completely and mop up the flood. Using a drain-and-trap auger long enough to reach the sewer outlet, probe and remove the obstruction, then flush with a hose. Coat the plug with grease or pipe compound and recap the cleanout.

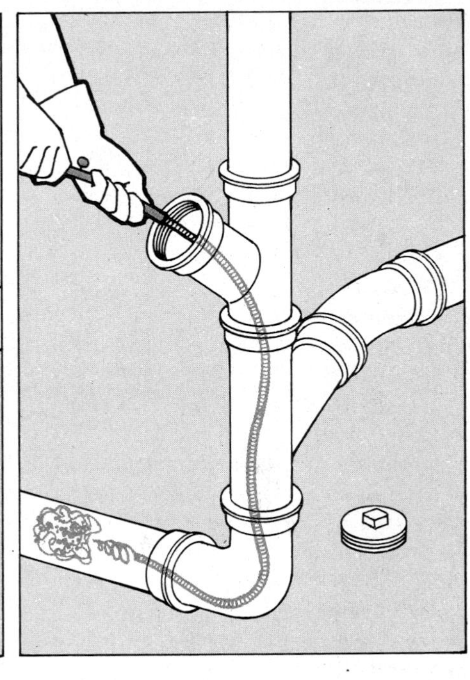

Getting at the Main Drain from the House Trap

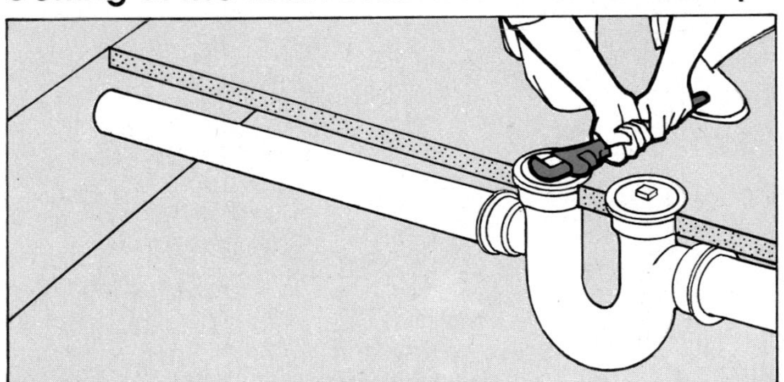

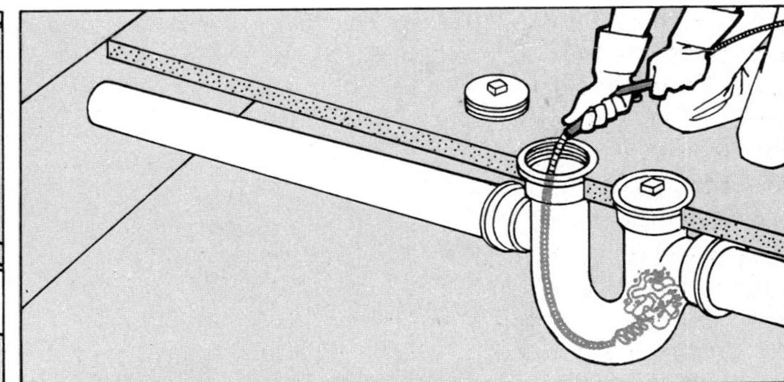

1 Opening the house trap. Locate this fitting by its two adjacent cleanout plugs, visible at floor level if the main drain runs under the floor. Spread heavy rags or stacks of newspapers around the trap to prepare for flooding, then slowly loosen the plug closest to the outside sewer line.

If no water leaks out as you unscrew the plug, the clog is in the trap or the main drain between the trap and main cleanout and may be fairly easy to remove *(Step 2)*. If water seeps out, probe the drain beyond the house trap with an auger. Unless you can remove such a blockage quickly, recap the plug and call the plumber.

2 Working from the trap. Unscrew the trap plug completely and feed an auger through the trap toward the main cleanout, but probe gently. Do not attempt to free the blockage all at once, as you would when working from above, but poke a small hole in it. Wait for water to drain gradually, then break up the blockage with repeated jabs of the auger. After the flow subsides, open both house trap cleanouts and scrape out any remaining sludge with a wire brush. (If the clog is not in the trap but in the main drain between the trap and the cleanout, remove the second trap plug and feed the auger toward the cleanout, following the same precautions.) Recap the house trap cleanouts and insert a hose into the main cleanout to flush the main drain. Replace the cleanout plug.

2 Big Savings from Small Jobs

A costly drip. As long as the faucet drips, the needle on the water meter turns, racking up a bigger bill for the homeowner. Stopping a drip from an old-fashioned stem faucet *(left)* is one job that most homeowners have learned to do. Fixing more sophisticated modern faucets is no more difficult—and prepares the owner for remedying a whole host of other plumbing problems, thus accomplishing a double saving: lower water bills and no cost at all for plumber's labor.

Almost any job seems difficult the first time you do it, whether it is making an omelet or putting a new cord on a lamp. The same is true for working on your house plumbing. A dripping faucet is a rudimentary problem that can be baffling; you must figure out how to get it apart—often the trickiest step of the operation—and what part is at fault. But once you have learned to cope with the everyday problems that beset sinks, toilets and tubs, you will save money and the inconvenience of waiting for a plumber to come. Some jobs in the bathroom or kitchen a plumber may not do at all, such as replacing a broken tile or a cracked soap dish. Many other repairs are so simple that today's highly paid master plumber is reluctant to take them on; his time is more profitably spent on large projects.

Even those who think of themselves as all thumbs can master the steps involved in basic plumbing repairs. There are a few elementary guidelines that make the task easier.

☐ Work slowly and be patient.

☐ Turn off the water supply. If you are working on a faucet, close the drain to keep screws and other small parts from falling in.

☐ When you dismantle the parts of a fixture, line them up in the order and orientation of disassembly so you can put them back together more easily without wondering which way a part faces.

☐ Do not force a part "frozen" by corrosion; apply a few drops of penetrating oil, wait a while—overnight if necessary—and try again.

☐ Inspect parts for signs of wear or corrosion while you have a fitting disassembled. Replacing a worn part avoids future trouble and the necessity of dismantling the fixture or fitting again.

☐ Keep on hand a supply of common replacement parts—faucet cartridges, washers, washer screws and O-rings. That way you avoid the bother of a special trip to the store and the nuisance of having to leave the water shut off while you are out matching the faulty part.

Finding the right replacement parts may be the most troublesome part of a plumbing repair. The best source is a plumbing-supply store, which will have a larger stock of fittings and more knowledgeable salespeople than a hardware store or housewares store. Generally it is better to buy parts made by the manufacturer of your fixture; in many cases it is essential, for no others will fit.

Be prepared for frustration in the search for replacement parts. There is no standardization in fittings and fixtures, as there is in piping; many manufacturers make several types of faucets, all different. And since such equipment may last for decades, you are likely to find yourself with a model for which parts are no longer available. If you cannot find a needed part, replace the fitting—the job is surprisingly simple and you gain in convenience and appearance.

Faucet Repairs: The Right Method for Every Model

A dripping faucet is not only annoying but stains and erodes a sink or lavatory and wastes water—25 drops a minute may consume one gallon a day. Although faucets come in a bewildering variety of sizes and shapes, there are actually only two basic types: stem faucets and single-lever faucets.

Single-lever faucets—once used mainly in kitchens but now popular in bathrooms because of the convenience of controlling the water and getting the right temperature with a single flick of the wrist—require repair less frequently. Water from both hot and cold supply pipes is controlled, in most types, by a single cartridge set into a chamber, where hot and cold water supplies mix while passing through the faucet. If a drip does occur, you replace the cartridge as described on pages 40-41.

A stem faucet is a threaded rod that fits into the port attached to the supply pipe—one faucet for hot, one for cold. If hot and cold mix, they do so in a shared spout, after passing through the faucets. Stem faucets are prone to leaks. Most often water drips from the spout because the seat washer or its mating metal seat wears, preventing a tight seal at the inlet in the off position. But water can also leak from the faucet body when the faucet is turned on—some of the flow to the spout escapes through materials that are supposed to seal the hole for the handle. Such defects are easily repaired by replacing washers or, if metal parts are badly worn, the stem and seat. In fact, the trickiest part of the job may be in figuring out how to take the faucet apart.

Before starting any faucet repair, turn off the water supply at the valve that is closest to the faucet—usually it will be underneath the fixture. Then turn the faucet on to allow any remaining water to run out. To prevent a pipe wrench from marring the chrome of the faucet, line the jaws of the wrench with electrician's tape. Or you can use a smooth-jawed monkey wrench.

Plug the sink drain so that small parts cannot fall down it, and line the sink with a towel to prevent damage from accidentally dropped parts or tools. Place the parts beside the sink in the exact order in which they were removed to facilitate reassembly.

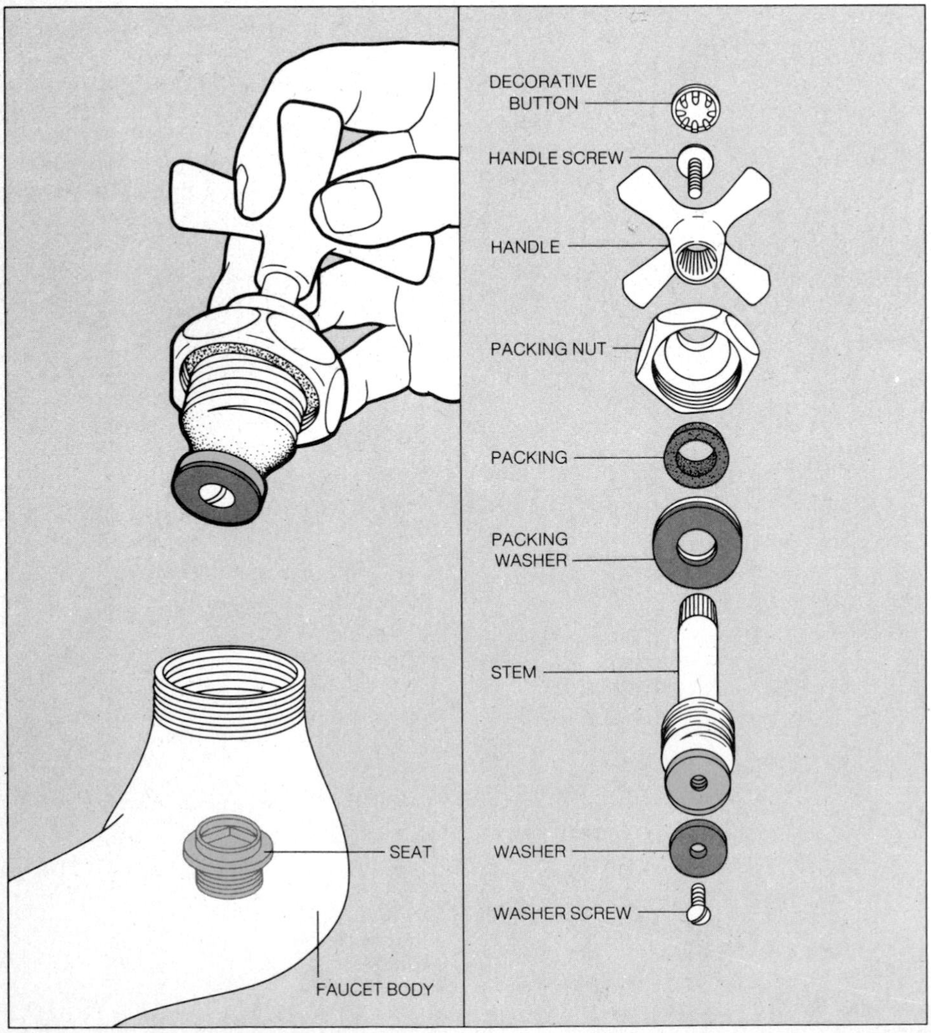

Anatomy of a stem faucet. A number of metal parts (generally brass) together with sealing materials of fiber, rubber or plastic make up the stem, which screws into the faucet body. The washer on the end of the stem seals against the seat, and the packing nut closes off the body. Leaks occur at the packing nut and seat. A decorative button generally covers the screw fastening the handle to the stem.

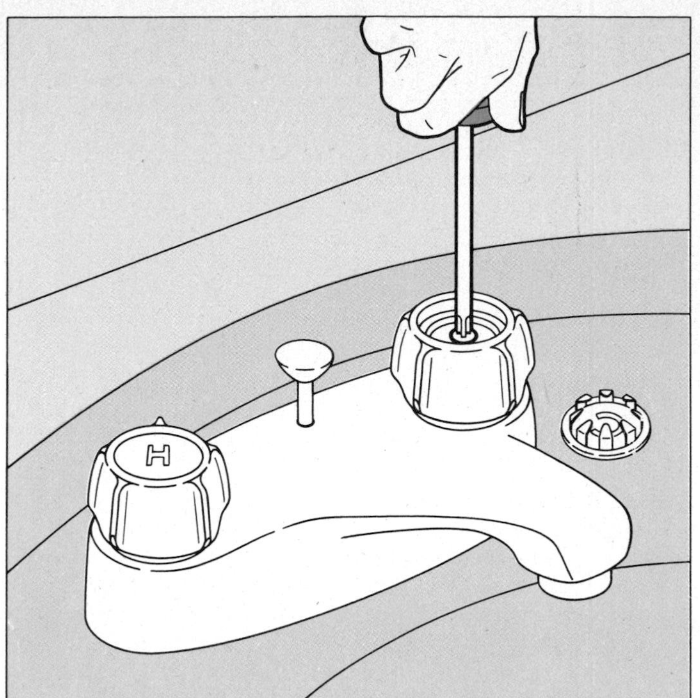

Taking a Stem Faucet Apart

1 Removing the handle. On some stem faucets the handle blocks access to the packing nut and must be taken off first. The screw in the top of the handle may be hidden by a decorative button. Use the point of a utility knife to pry it out. Remove the handle screw and pull the handle straight up. On older faucets like the one on the opposite page, the packing nut is directly accessible, and you need not remove the handle from the stem when replacing a washer.

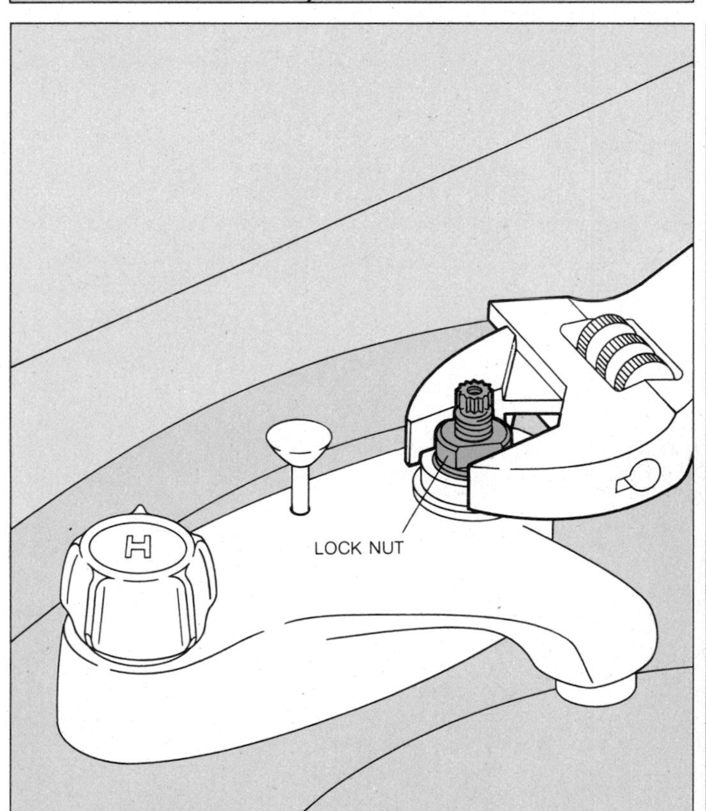

2 Removing the locknut. In this type of faucet a locknut serves as a packing nut; turn the faucet to a half-open position and unscrew the locknut counterclockwise with a wrench. The stem may come out with the locknut. To separate them, protect the stem with electrician's tape, clamp it in a vise and twist the locknut off with a wrench.

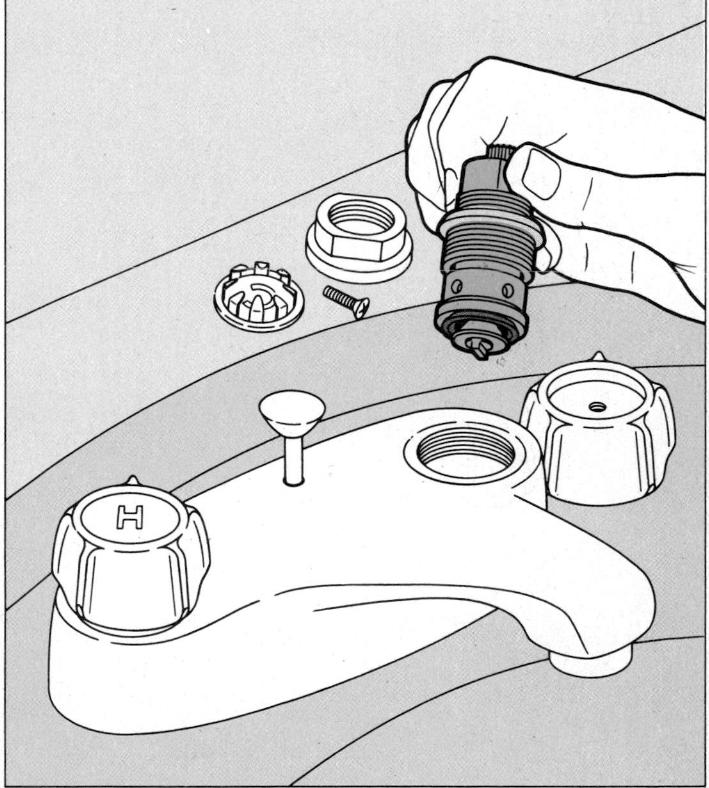

3 Taking out the stem. If the stem does not unscrew easily, replace the handle and turn it in the same direction that you would to turn on the water; keep turning until the stem comes out. If the faucet is fitted with a cartridge-type stem assembly *(page 33)*, pull it straight up by hand or protect the stem with cloth and lift it out with pliers.

Stopping Spout and Stem Drips

A steady slow dripping from the spout of a stem faucet signals the need for a new washer, a new seat or perhaps both. If the spout is a mixer that serves both faucets, you will have to test the drip to determine whether the hot or cold faucet is at fault.

In newer faucets with valve-stem assemblies or cartridges instead of a standard stem, O-rings or a rubber seat may take the place of a washer. Only after the faucet is disassembled can you determine what kind of stem faucet you have and what part you need to replace.

Finding a washer or O-ring of the right size and shape may be difficult. Replacements are generally sold in packaged assortments that may not contain any washers that fit exactly. With some judicious sanding *(below, center)* you can tailor the washers to the precise diameter that you need.

If the brass screw that holds the old washer in place cannot be turned, apply penetrating oil to it and wait 15 minutes to an hour before trying again. If the head of the screw breaks off, pry out the old washer and use pliers to remove the screw. When changing the washer, replace a corroded or damaged screw.

A faucet seat that is marred will quickly wear out a new washer, so while the faucet is disassembled, inspect the seat for scratches and pitting *(page 34)*. If you find signs of wear or if a new washer fails to stop the faucet from dripping, replace the seat with a new one or dress it with a seat-dressing tool.

Types of washers. Standard stem-faucet washers may be flat or beveled with a hole in the middle for the washer screw. One type of faucet uses a washer-like diaphragm that covers the entire bottom end of the stem. Other stems are fitted with a seat ring that acts like a washer. Cartridge-type stem faucets may have a washer and spring that are seated in the faucet body.

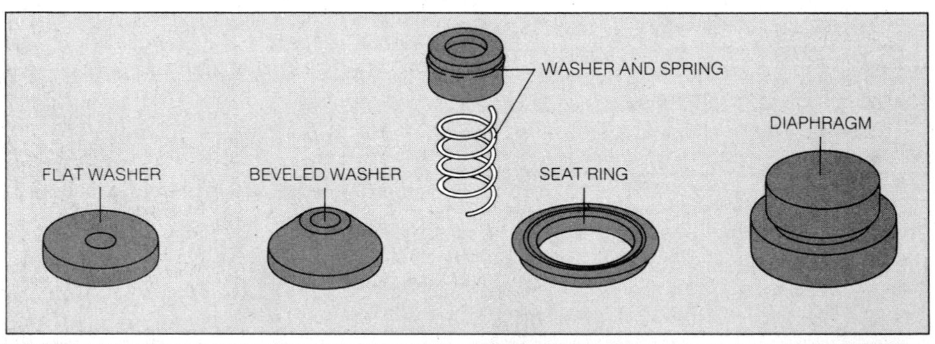

Sanding a washer to size. If you cannot find a washer of the right size, choose one slightly larger than the old one. Tape a piece of very fine sandpaper to the workbench. Push the washer over a drill bit that will hold the washer without letting it slip and chuck the bit in an electric drill, preferably of variable speed. Turn on the drill and, using very light pressure, press the edge of the spinning washer against the sandpaper to reduce the washer evenly all around.

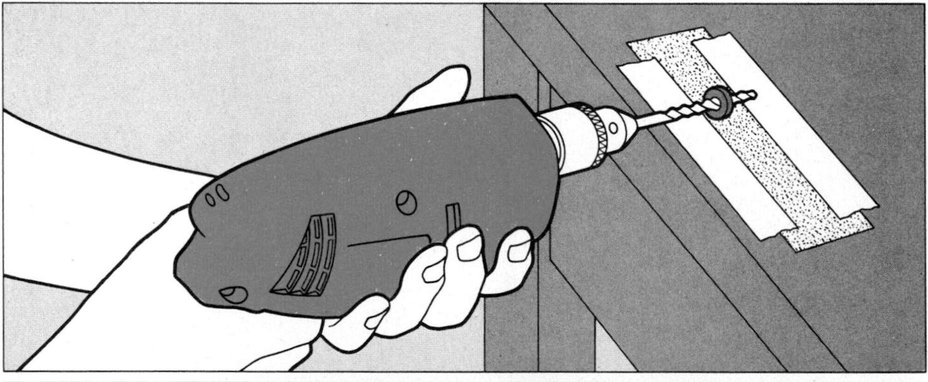

Repairing a diaphragm faucet. Suction of the diaphragm in the valve seat may make this stem assembly difficult to remove from the faucet body. Wrap the top of the stem with cloth and pull out the stem with pliers. If the old diaphragm sticks inside the faucet body, pry it out with the tip of a screwdriver. With a flashlight, inspect the faucet body to make sure that there are no pieces of the old diaphragm remaining inside it; otherwise, the new one will not seat properly. Fit the new diaphragm over the bottom of the stem, making sure the diaphragm is snug all around.

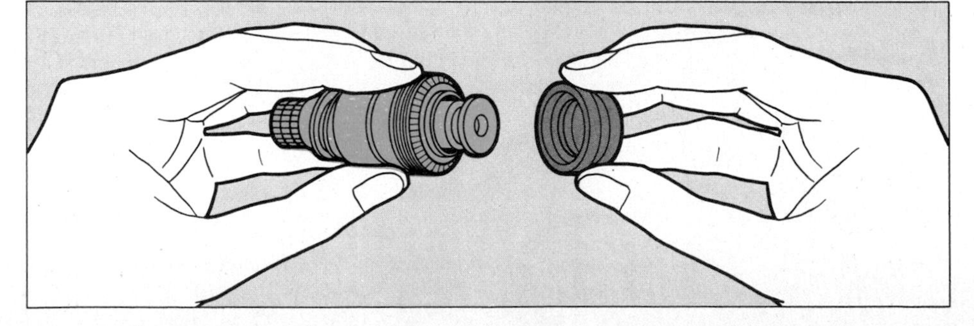

Repairing a reverse-pressure faucet. In this arrangement, the faucet seat is built into the stem rather than the faucet body. When the faucet handle is turned to the off position, the washer moves upward to close against the seat, opposite to the direction taken by the washer in a standard stem faucet. To replace the washer, unscrew the stem nut and remove it along with the metal washer, rubber washer and washer retainer. Pull off the metal seat and inspect it; replace it if the surfaces are worn. Press the new washer, beveled side up, into the retainer cup.

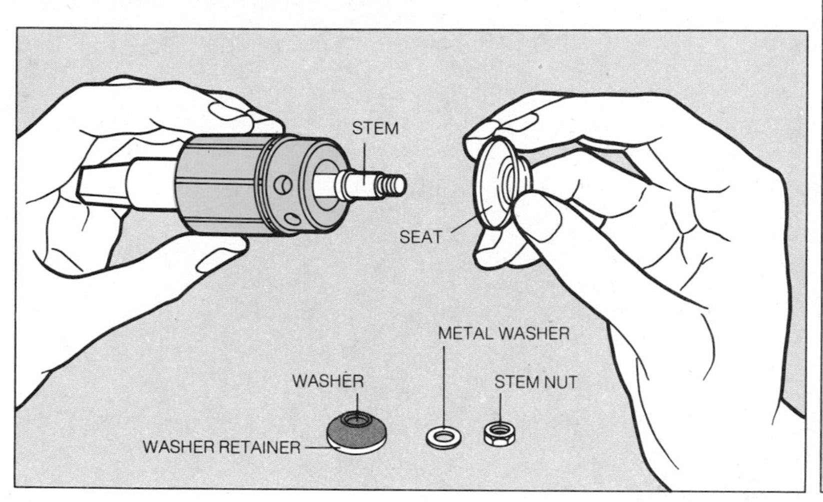

Repairing a seat-ring faucet. With this type of stem, hold the rectangular end of the stem with long-nose pliers and unscrew the threaded centerpiece, which holds the seat ring used instead of a washer. Slip off the sleeve and replace the seat ring, making sure that the lettering on the new ring faces the threaded part of the stem.

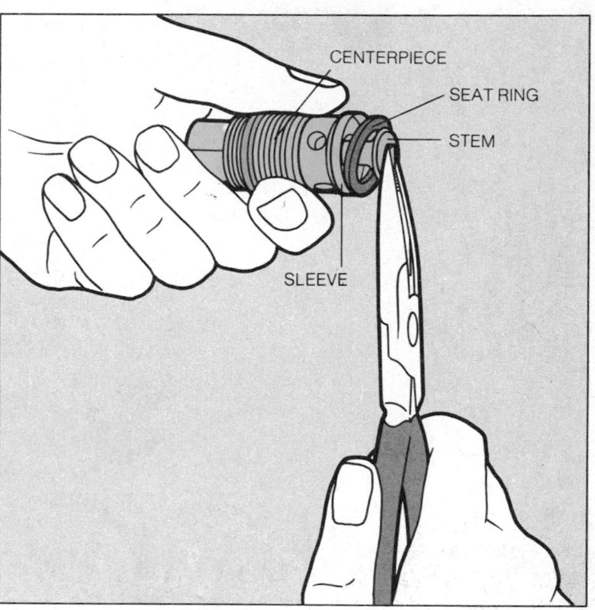

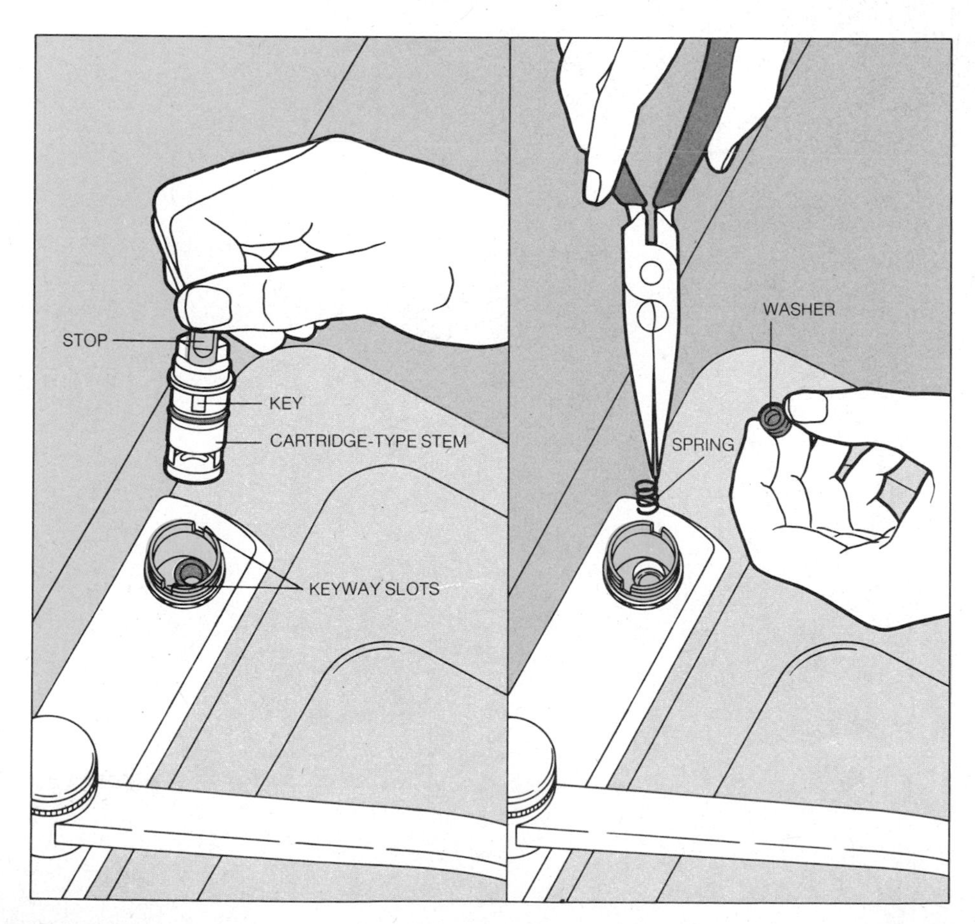

Repairing a cartridge-type stem faucet. In this faucet a cartridge takes the place of the stem; the cartridge presses against a separate washer and spring in the faucet body. Usually only the washer and spring require replacing. Lift the cartridge out of the faucet *(far left)*, making sure to note the correct alignment of the stop on the top of the cartridge and the keys on its side that fit into the two keyway slots on the faucet body. With long-nose pliers pull the rubber washer and metal spring out of the faucet body *(left)*. Push the new washer firmly into place with your index finger. If the spout still drips, replace the cartridge as well as the washer and spring.

Renewing Valve Seats

Inspecting the seat. If the spout continues to drip after you have replaced the washer, look at the valve seat for signs of wear—scratches, pits or an uneven surface. Use a flashlight to illuminate the inside of the faucet and, as a further check, run the tip of a finger around the edge of the seat.

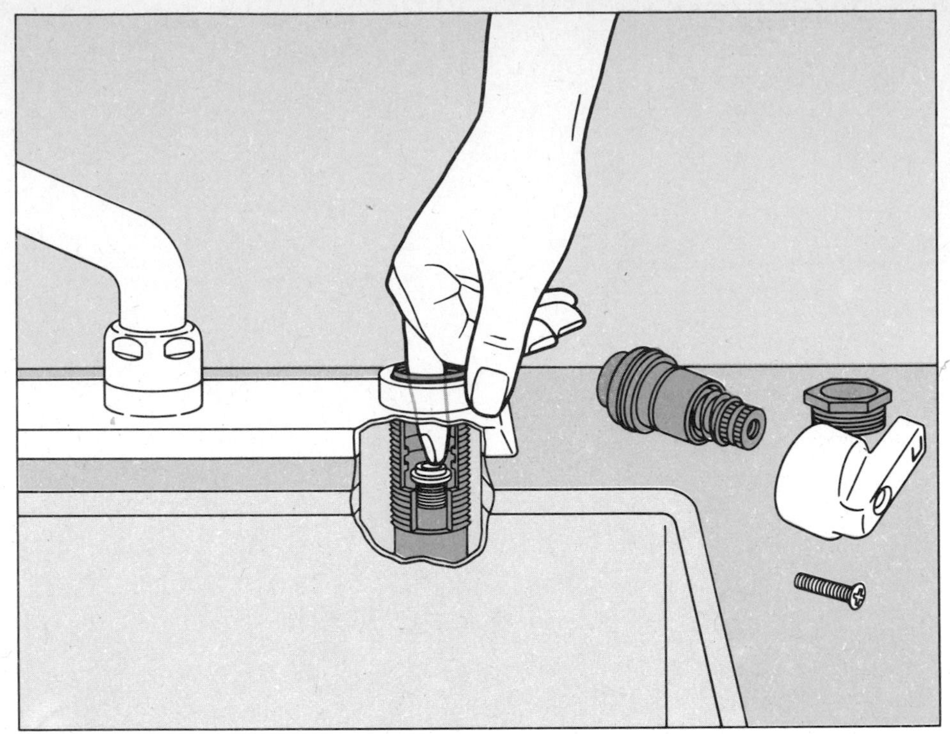

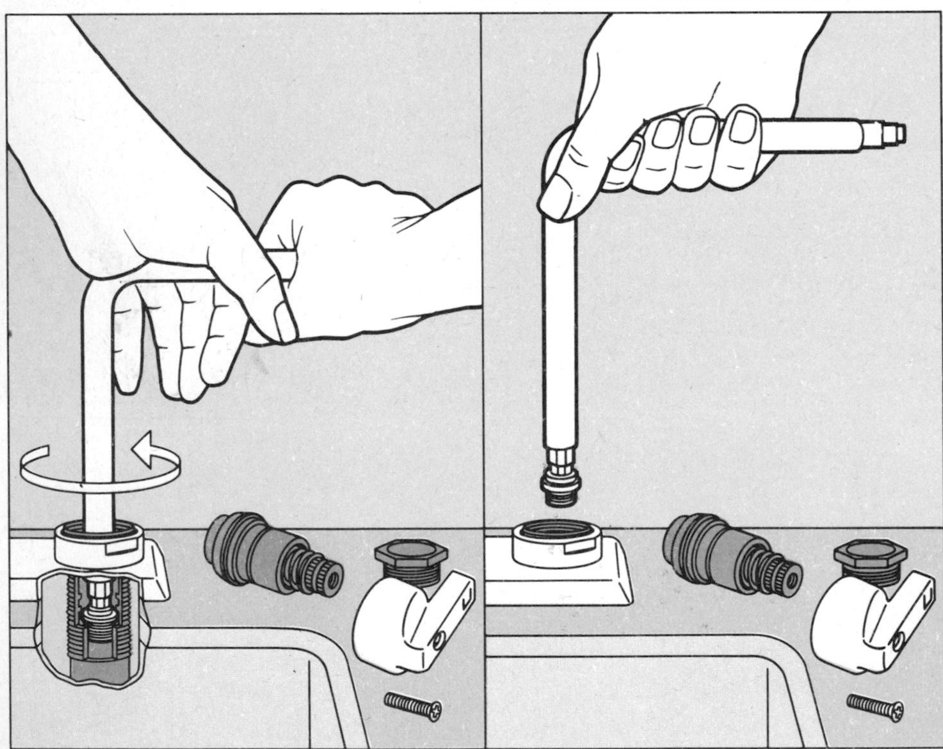

Installing a new seat. Most faucet seats are removable, and it is better to replace a damaged seat than to try to smooth a worn one. Using a seat wrench, turn the seat counterclockwise and lift it out *(above)*. Take it to your plumbing-supply store to get an exact duplicate. Lubricate the outside of the replacement with a pipe-joint compound, push it firmly onto the seat wrench and screw it into the faucet body *(center)*.

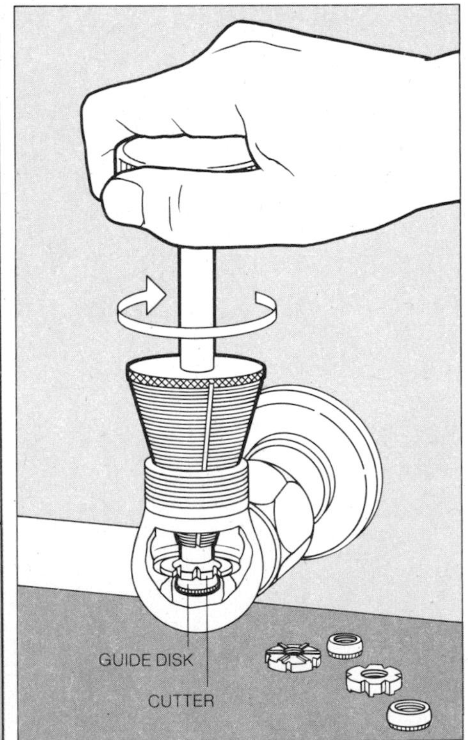

Dressing a seat. If a worn valve seat cannot be removed—it may be built into the faucet—its surface must be ground smooth with a valve-seat dresser. Use the largest cutter that fits the faucet body. Screw on a guide disk that just fits the valve-seat hole. Slide the cone down snugly into the threads of the faucet body. Pressing down lightly, turn the handle clockwise several times. Remove the filings with a damp cloth.

Three Ways to Seal a Stem

Replacing a packing washer. A dribble around the stem of a faucet when the water is running can often be stopped simply by tightening the packing nut—but do not overtighten it.
If the stem leak persists, the packing should be renewed. Packing comes either as graphite washers or as self-forming packing, a graphite impregnated twine. Some modern faucets use O-rings or cork gaskets for packing.

To replace a packing washer, remove the handle and packing nut, clean out the old packing and slip on the new. For an old faucet that used self-forming packing, you may find a ready-made packing washer that fits. Push the packing washer onto the stem as far as it will go, and screw the nut over it, turning it clockwise.

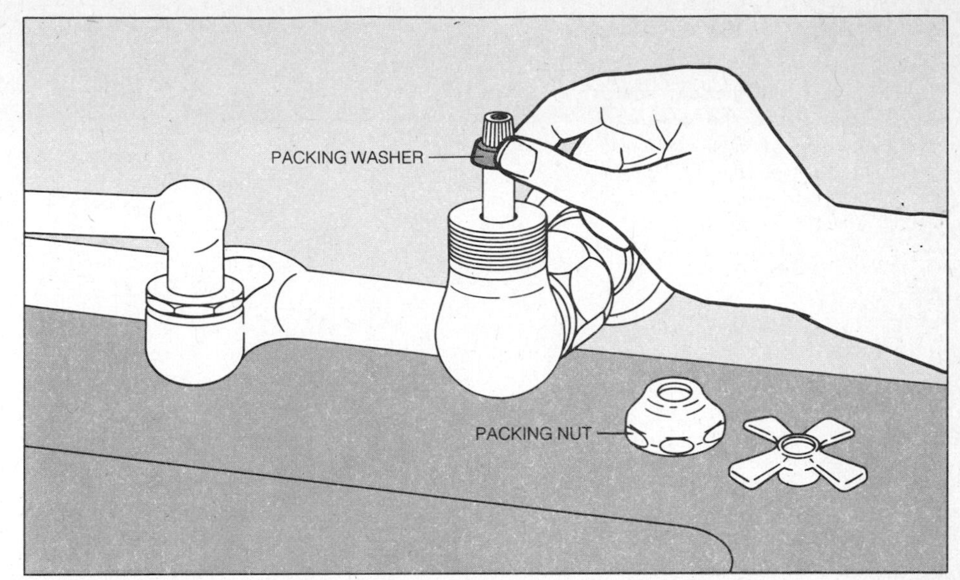

Using self-forming packing. Untwist the strands making up the spooled twine and use one at a time. Wrap it in layers around the faucet stem, using half again as much material as would be needed to fill the packing nut. Then reassemble the faucet. The packing compresses into solid form when the nut is screwed down over it. Self-forming packing can be used instead of a packing washer if you cannot find one to fit.

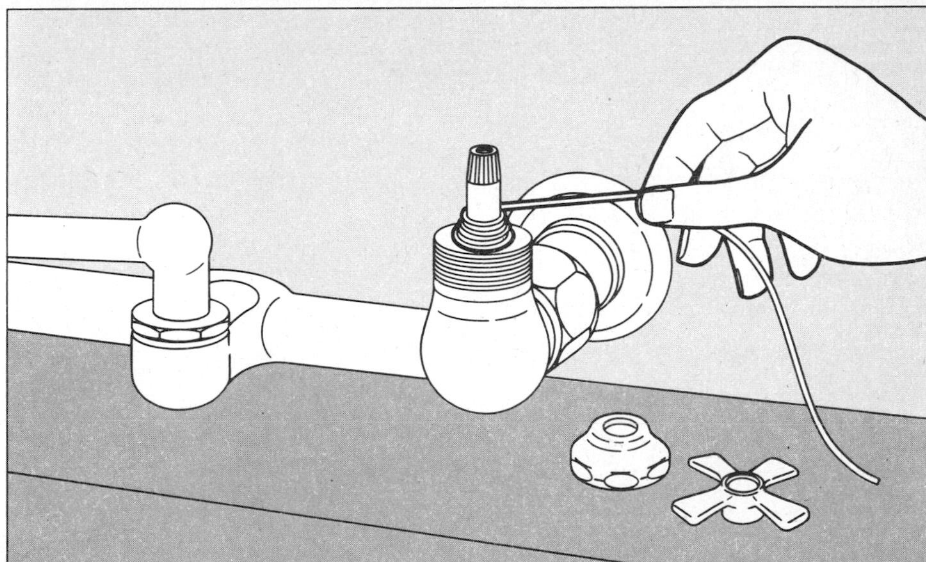

Replacing O-rings. Pinch the O-ring with your fingers to make a loop; then grasp the loop and pull the ring off. There may be more than one on a valve-stem assembly like the unit illustrated —one outside the lock nut, the other uncovered when the lock nut is unscrewed.

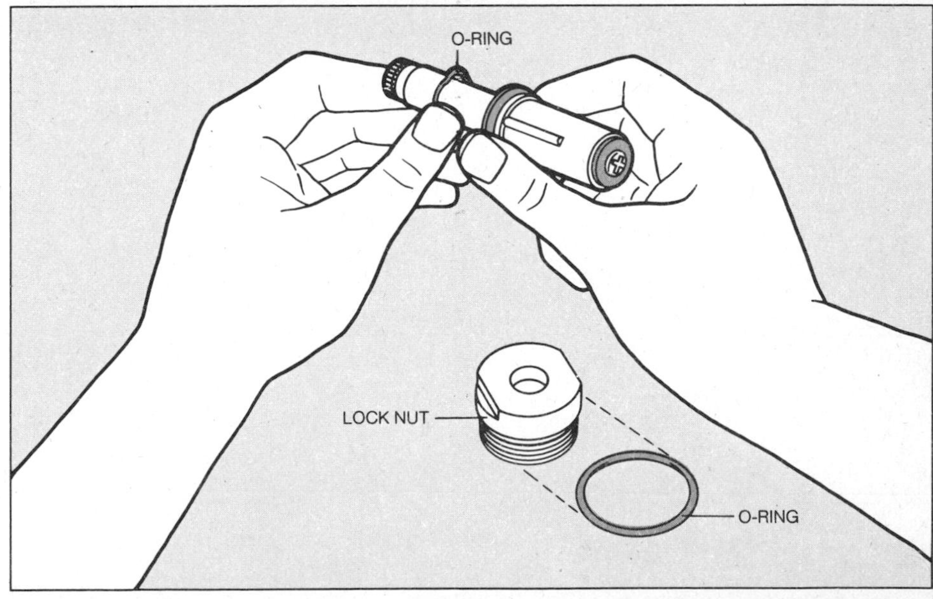

A Freezeproof Outdoor Faucet

A drip from a freezeproof outdoor faucet is repaired just like one from any stem faucet, by replacing the washer *(page 32)*. The difference is in taking apart one of these devices, specially built to prevent freezing in sub-zero weather. The packing nut and stem require several extra steps to disassemble *(below)*.

Drips are a common problem with these faucets because their design leads many people to turn the faucet off too hard, causing unnecessary wear on the stem washer. The faucet is installed *(page 110)* at a slight tilt toward the outside wall, so that when the handle is turned off, the small amount of water remaining in the faucet body continues to run out of the spout until the body is completely empty. This draining action makes the faucet freezeproof, of course, but instead of waiting a minute for the trickle to stop, homeowners unfamiliar with the mechanism often try to turn off the faucet harder and thus wear out the washer.

Inside a freezeproof faucet. All the parts of a standard stem faucet are present in a freezeproof faucet, but the sloped, elongated body of the faucet allows the stem to stop the flow of water inside the house, where the temperature stays above the freezing point. Water remaining in the exposed body drains out of the spout.

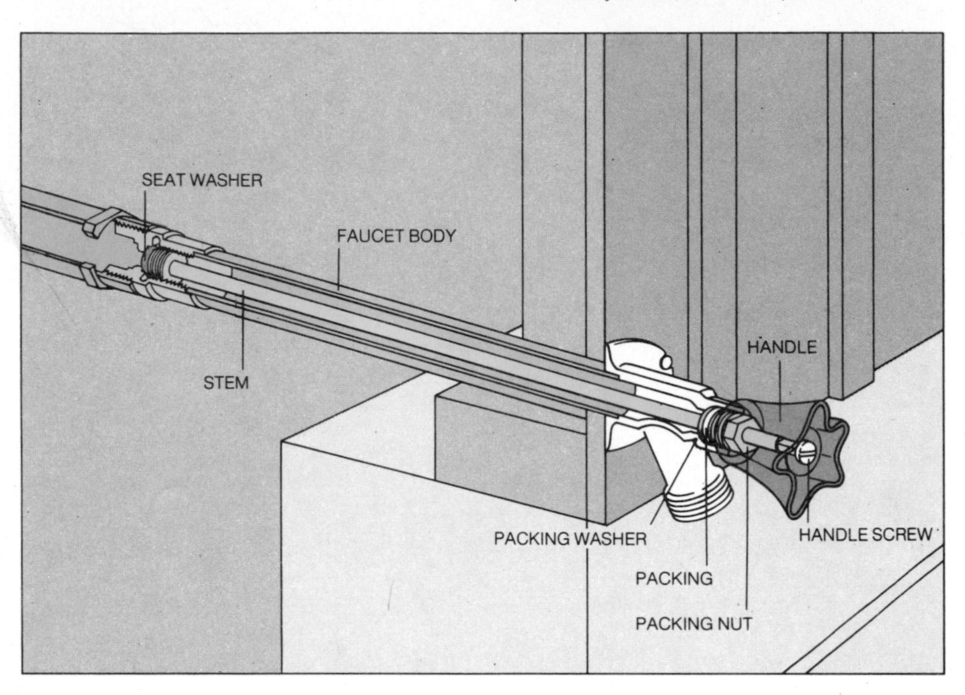

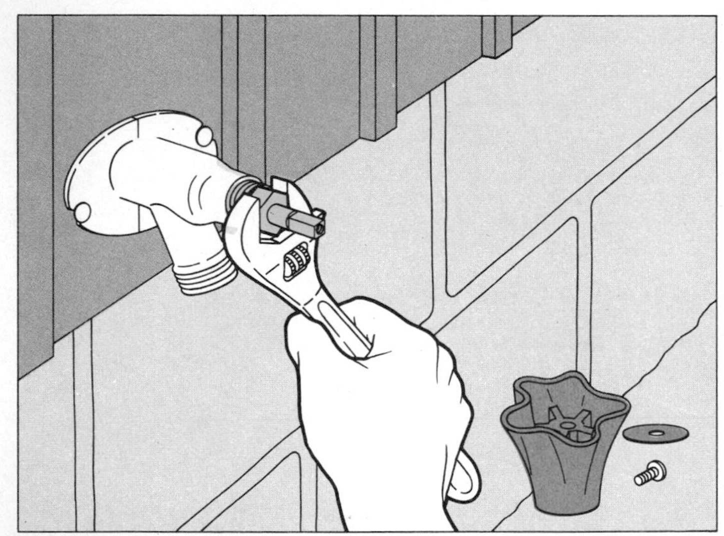

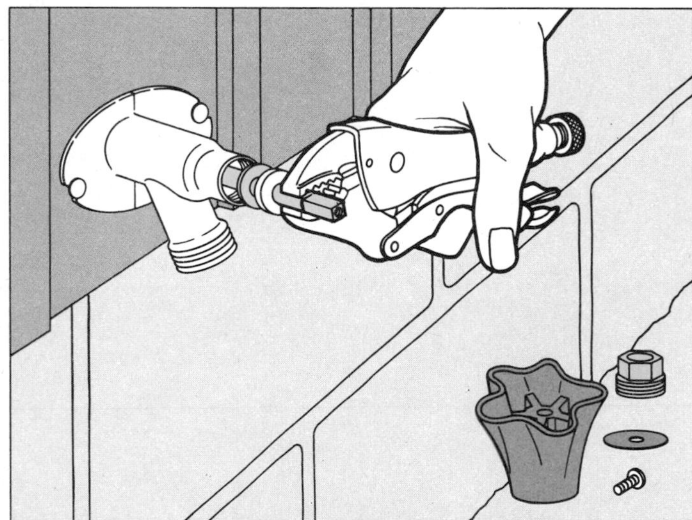

Tricky Technique for an Elusive Washer

1 Removing the packing nut. Remove the handle screw and the handle. Unscrew the hexagonal packing nut—it is designed to come off, even though the faucet body may appear to be all of one piece with the stem projecting from it. Then put the handle back on the stem.

2 Removing the stem. Turn the handle counterclockwise to unscrew the washer end of the stem from the faucet body. Then pull the handle away from the faucet body. If the stem cannot be budged this way—the packing holds it very tight—remove the handle, set a pair of locking-grip pliers over the round part of the stem and pull the stem free. Then replace the stem washer in the ordinary way.

Single-Lever Faucets

Kitchen or bathroom faucets with a single handle to control the flow and mix of water generally serve for years without requiring attention. When they do develop drips or leaks, the entire control unit may have to be replaced.

As with stem faucets, the first problem is disassembly. Faucet makers conceal screws in mysterious places—under handles or decorative buttons—or eliminate them entirely. The drawings here show how to get inside four types of single-lever faucets. Once inside, you can see what kind of control unit the faucet uses.

It is important to get identical replacement parts, as each company makes its own. Consult the telephone directory for the maker's nearest sales office; it will direct you to a local parts distributor.

Disassemble these faucets with care, noting how parts fit together—and in what order—as an aid to reassembling them. Sketches can be very helpful as an aid for reassembly.

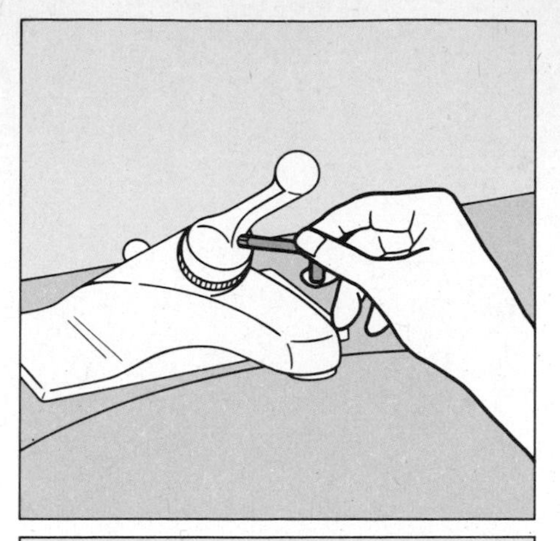

Ball faucet. As the round shape of the faucet handle suggests, the control inside this faucet is a ball made of brass, with three ports through which water flows when the ports are centered over seats in the faucet body. Under the shank of the handle is a setscrew that must be loosened with a hex wrench to get the handle off. Do not take the screw all the way out; it is easily lost.

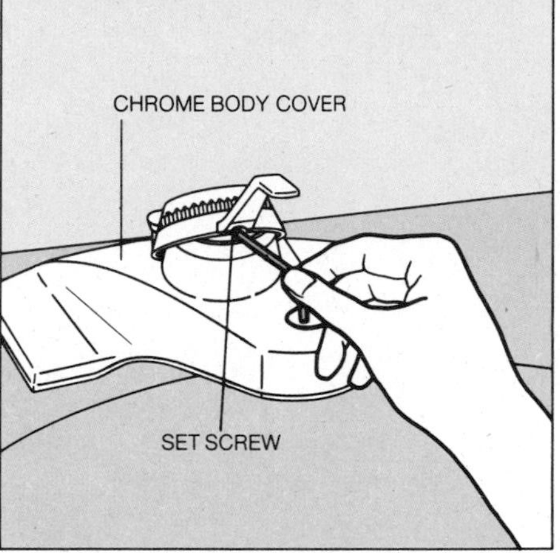

Ceramic-disk cartridge faucet. The screw that releases this handle is not beneath the decorative knob where you might expect to find it. Raise the lever as high as it can go; recessed under it is a small setscrew that releases the handle. But the control unit is now only partially uncovered; you must also remove the chrome body cover. To do this, disconnect the thumbscrew holding the pop-up drain rod beneath the sink (first mark the rod with a felt-tipped pen so you can connect it in the same place). Then unscrew the two Phillips-head screws on the underside of the faucet and lift off the body cover. Newer versions have a slot screw in the handle and a chrome ring that unscrews for easier access to the cartridge.

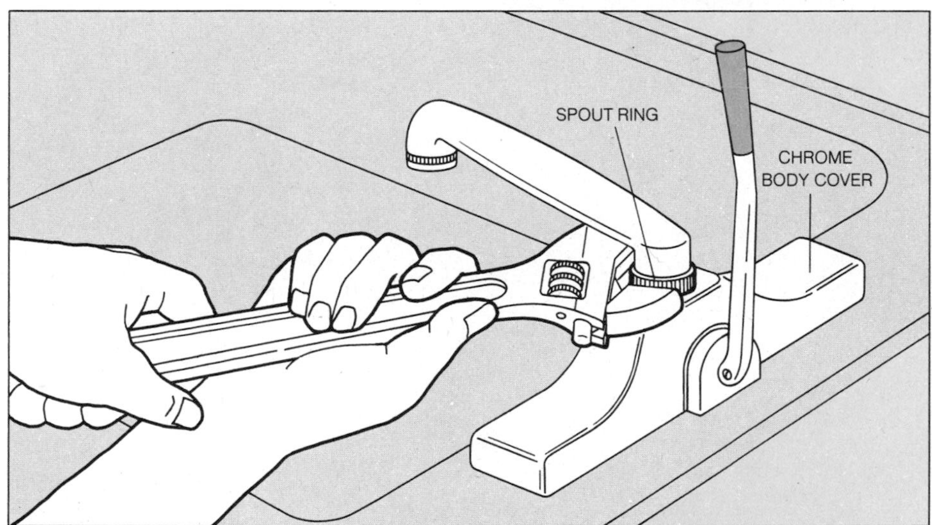

Tipping-valve faucet. So-called because the hot- and cold-water inlets each have a valve that tips when the lever is moved forward, allowing water to pass through to the spout, this faucet is widely used in kitchens, although it is no longer made. To get into the faucet, turn the spout ring counterclockwise with a wrench (if you use a pipe wrench, be sure the wrench's jaws are covered with electrician's tape) until the spout can be lifted off. Then pry up the chrome body cover with a screwdriver and lift it off.

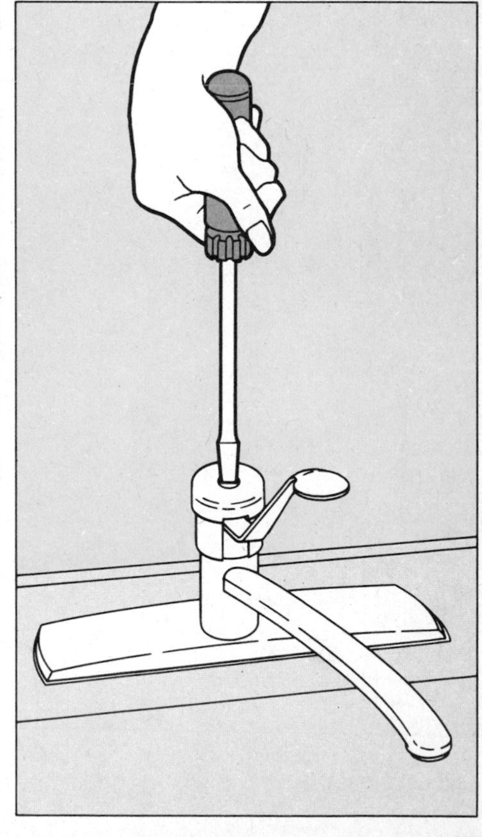

Sleeve-type cartridge faucet. The screw in the cover of the faucet handle illustrated is in plain sight, but in newer models it is hidden by a decorative button that must be pried off. The faucet lever raises to turn the water on and lowers to turn it off. Inside the cover the motion of the lever moves a cartridge up and down to turn the water off and on. An old model of this faucet has a retainer clip in the outside of the handle that must be pulled out for disassembly.

A Tipping Valve

1 Removing the strainer plugs. Before concluding that the parts of this faucet *(inset)* need replacing, try curing a drip from the spout or sluggish flow of water by flushing dirt from the faucet. Turn the water on hard and move the lever from left to right several times. If that does not work, remove the spout and chrome body cover *(page 37)* and unscrew the strainer plugs.

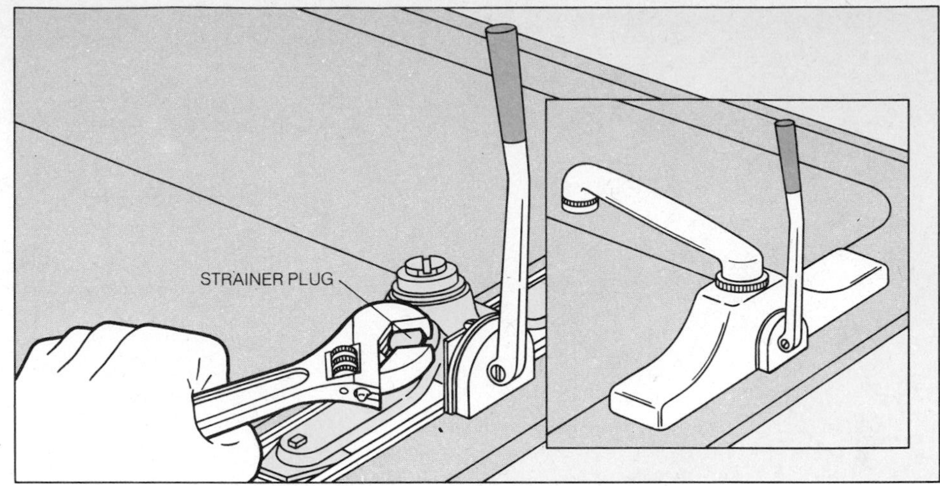

2 Cleaning the valve assembly. One by one take out the valve parts, using a seat wrench to remove the valve seat *(page 34)*. Inspect the faucet seat and the valve stem; any signs of wear indicate that all parts should be replaced. Otherwise, clean the parts with an old toothbrush and soapy water. Replace the parts in the order shown, and try the faucet again. If dripping persists or if any of the parts show signs of wear, buy a kit to replace all parts except the strainer and plug in both sides of the faucet.

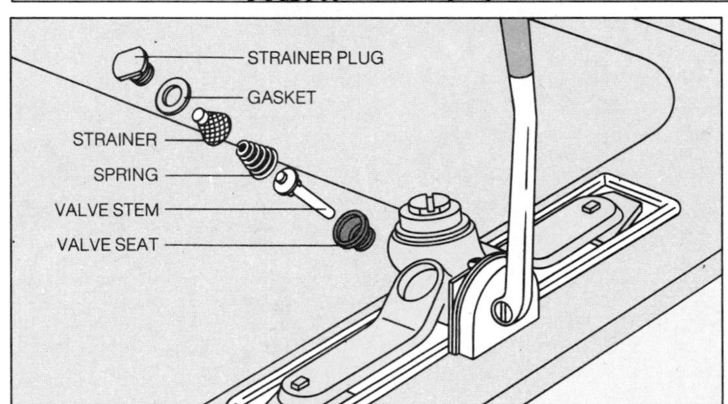

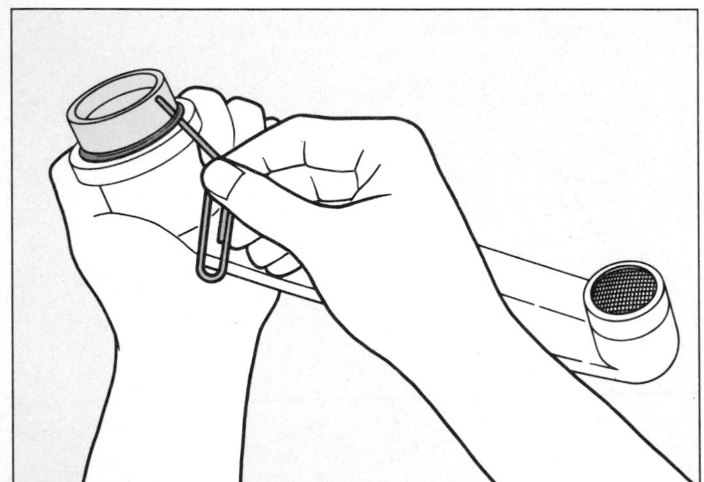

Replacing the spout O-ring. Like many other faucets with a mixing spout that swivels from one side to the other, this one has a rubber O-ring that prevents water from dribbling out where the spout joins the faucet body when the faucet is turned on. To replace the O-ring, pry it off with a straightened paper clip—but be careful not to scratch the metal. Take the O-ring to the store to get an identical one. Roll the new one into place and lubricate it with O-ring lubricant. If a new O-ring fails to stop the water from dribbling, replace the entire spout assembly.

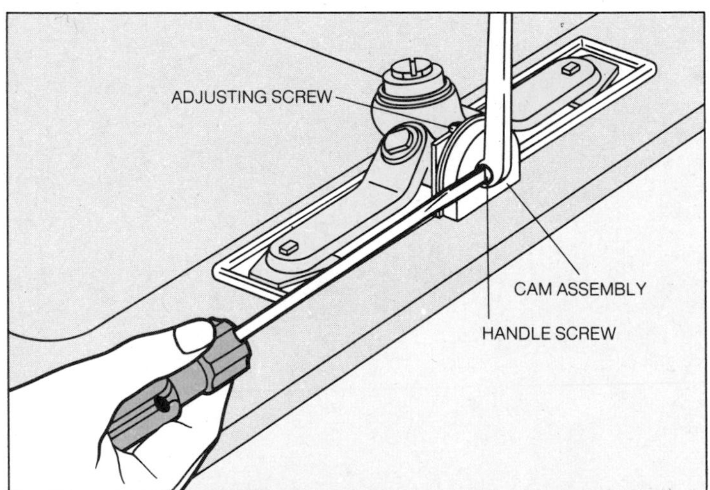

Fixing a loose handle. If the handle seems to work loosely, first tighten the screw that holds the handle to the cam assembly. (Newer models have two spring-steel pins that are force-fitted into the handle.) If, after tightening the handle screw, the handle still seems loose, remove the screw. If the unthreaded portion of the screw beneath the head is worn flat, replace it. If looseness persists, tighten the adjusting screw on top of the cam assembly about one quarter turn.

A Rotating Ball

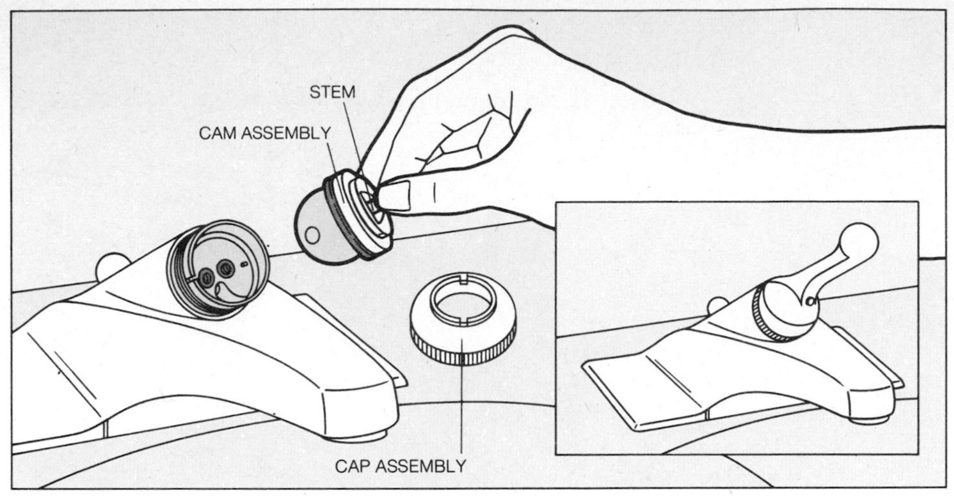

1 Removing the cap and ball. If the end of the spout drips when the water is off, replace the two rubber valve seats and steel springs in the bottom of the faucet body. To get at them, unscrew the cap assembly and lift out the stem of the ball; with it will come the plastic and rubber cam assembly.

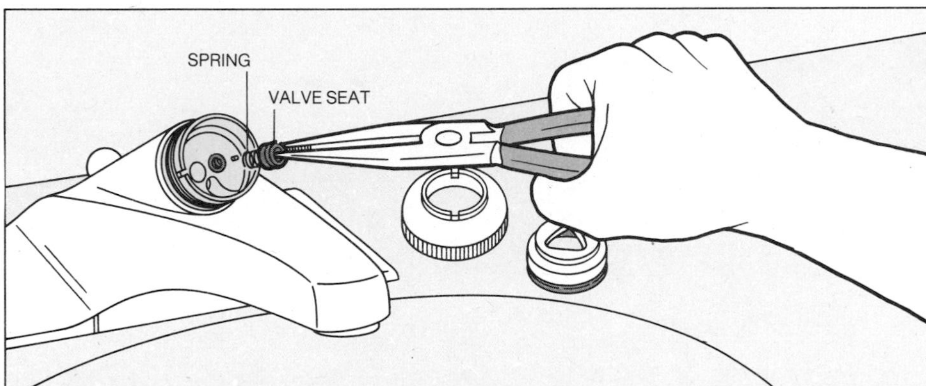

2 Replacing valve seats and springs. With long-nose pliers remove the valve seats and springs. Push replacements firmly into place with a fingertip. While you have the ball out, inspect it; if it is rough or corroded, replace it.

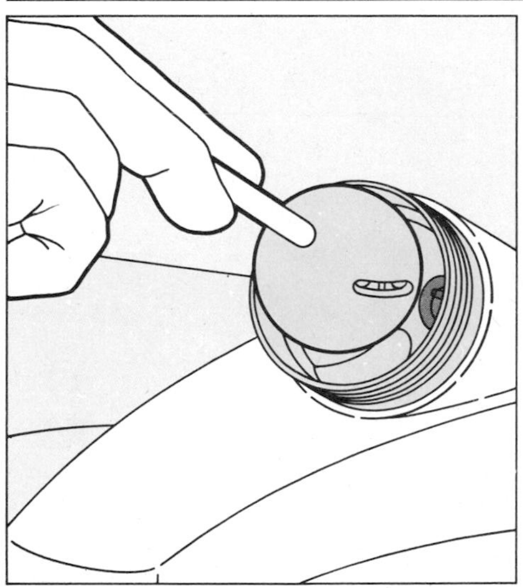

3 Replacing the ball. A tiny metal peg projects from one side of the cavity into which the ball fits. As you replace the ball, make sure that the peg fits into an oblong slot on the ball.

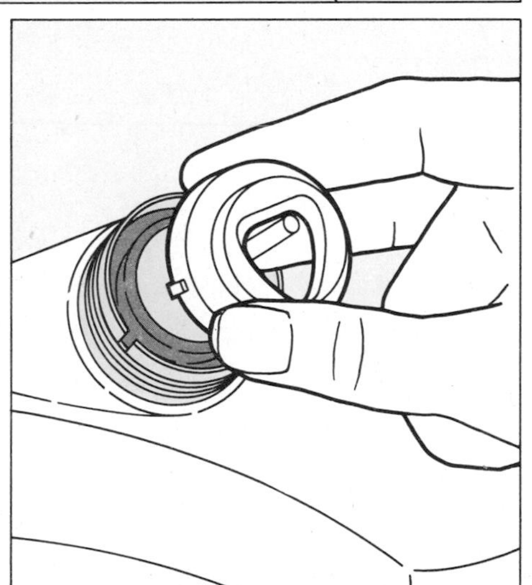

4 Replacing the cam assembly. On one side of the cam assembly is a small tab that fits into a slot on the faucet body. Replace the cam assembly as shown; screw on the cap assembly.

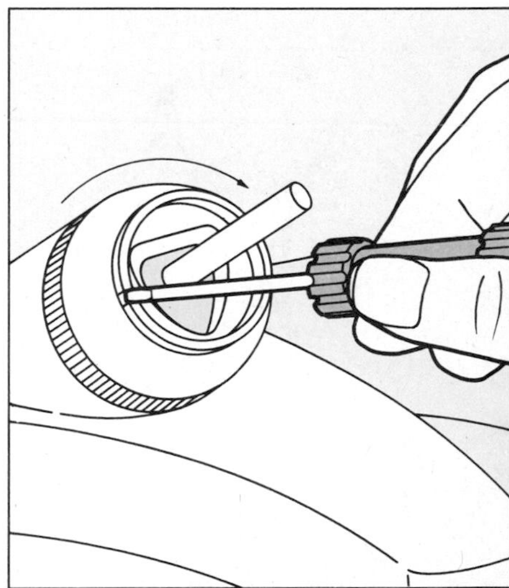

5 Setting the adjusting ring. Before reattaching the faucet handle, move the ball's stem to the on position. No water should leak out around the stem. If it does, with the special tool provided by the maker or the tip of a small screwdriver, tighten the adjusting ring by turning it clockwise. If, in order to stop the leaking, you have to tighten the ring so that the handle is difficult to work, then the cam assembly—both the plastic part and the rubber ring—need to be replaced. Position the handle so that the setscrew is over the flat on the stem, and tighten the screw.

A Metal Sleeve Cartridge

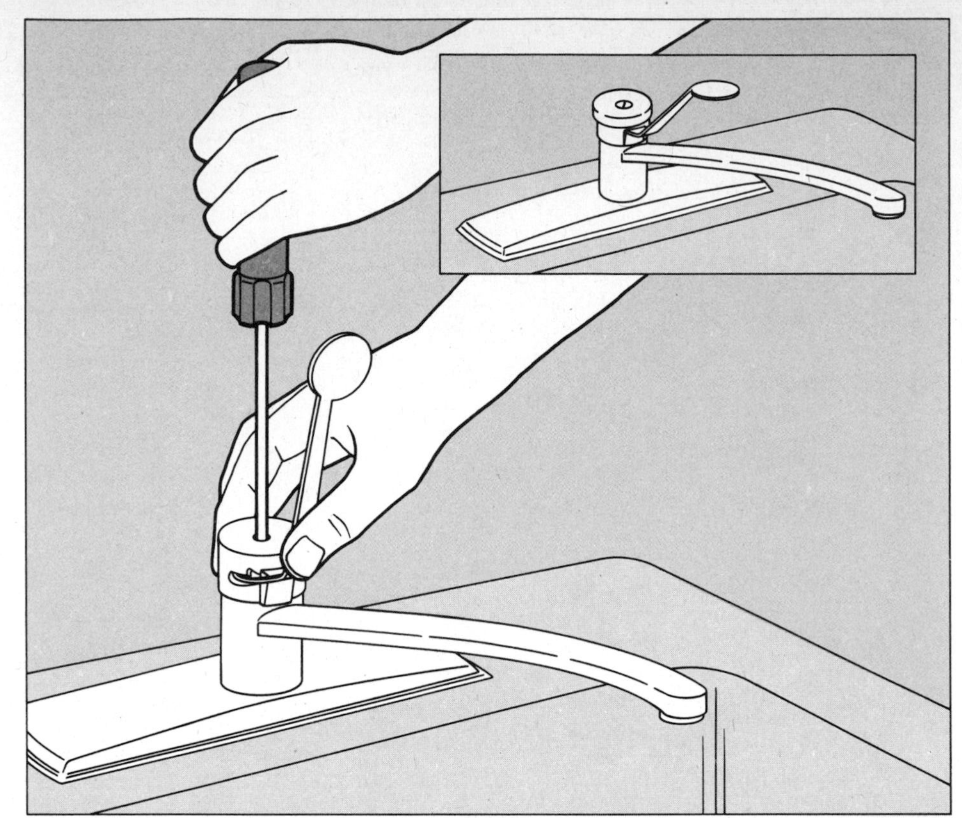

1 Removing the lever. After the lever cover is off *(page 37)*, push the lever down and place the tip of a screwdriver in the screw hole. With the screwdriver, hold down the cartridge stem so that it does not rise while you lift off the lever and its plastic housing.

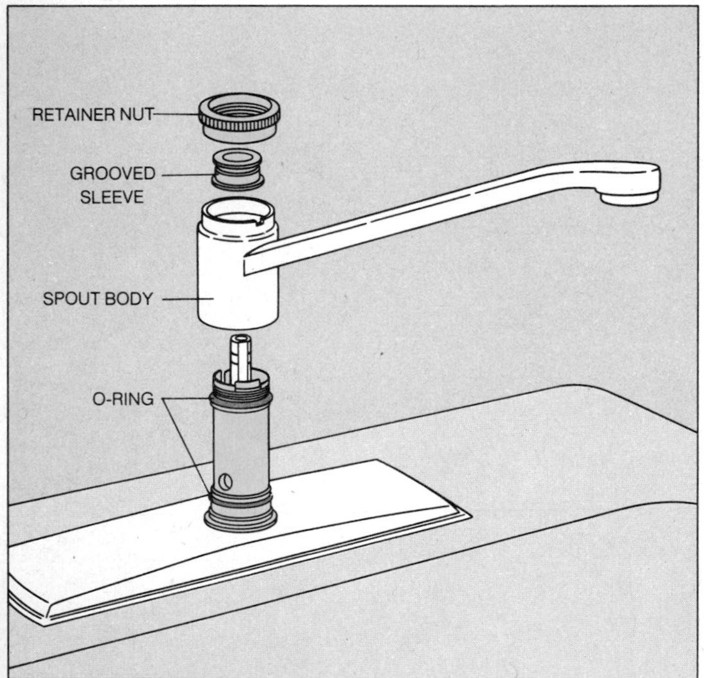

2 Removing the spout body. Unscrew the brass retainer nut and lift off the grooved sleeve. Lifting and twisting, pull the spout body off. Beneath it are two O-rings that prevent leaks around the body when the water is on. If the faucet has such a leak, replace the O-rings with identical ones. Put the spout, the grooved sleeve and the retainer nut back on, and replace the lever, following the procedure in Step 5.

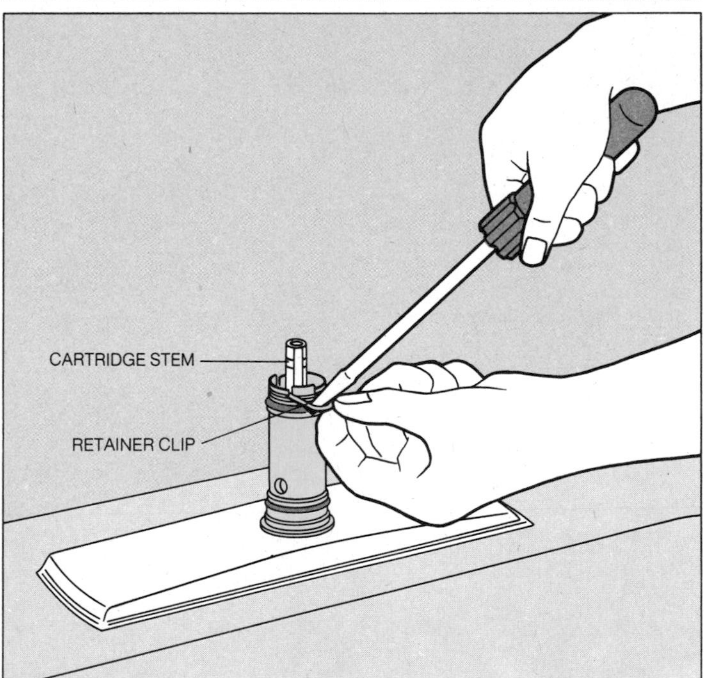

3 Removing the retainer clip. If your faucet drips slowly when the water is off, first make sure that the lever is engaged under the lip of the sleeve *(Step 5)*. If that does not cure the drip, remove the old cartridge and replace it with an identical one. Use a screwdriver tip to pull out the copper retainer clip that holds the cartridge in the faucet body. Then lift out the cartridge, using pliers to grip the top of the stem if necessary.

4 Reassembling the faucet. With the cartridge stem pulled up, push the cartridge by its ears down into the faucet body. The cartridge ears must be facing front and back, filling the space between the higher portions of the faucet body. One of the two flats on the cartridge stem is painted red; it must face toward the sink—otherwise the hot and cold water will be reversed. Slide the retainer clip in either side of the cartridge ears, making sure that it fits tightly. Replace the spout and the grooved sleeve, and tighten down the retainer nut.

5 Replacing the lever. Pull the cartridge stem up as high as it will go, using pliers if necessary. Raise the lever to the on position, place it over the faucet assembly and engage the ring inside the lever housing under the lip of the grooved sleeve. Move the lever gently from side to side until the housing drops into position. Replace the lever cover, and tighten the screw.

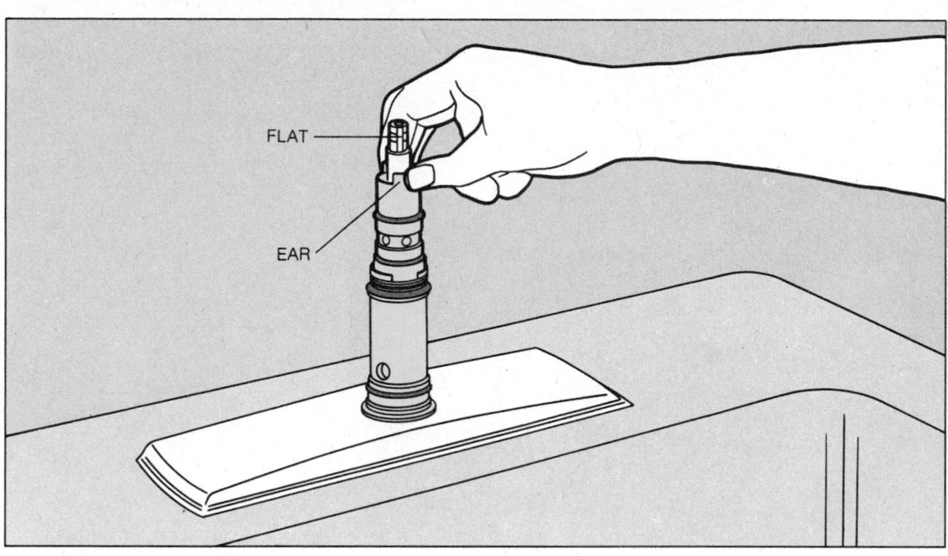

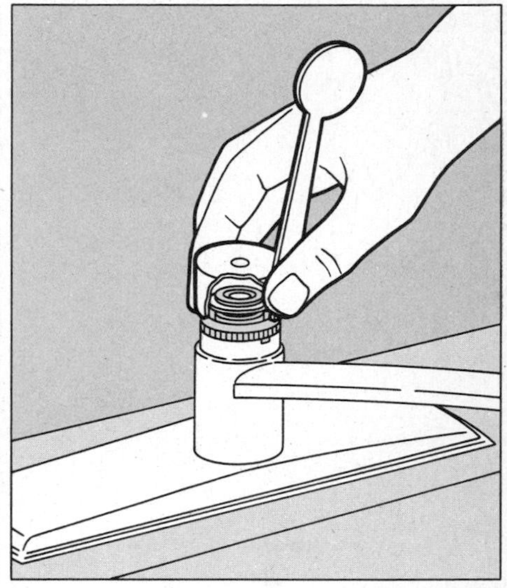

A Ceramic Disk

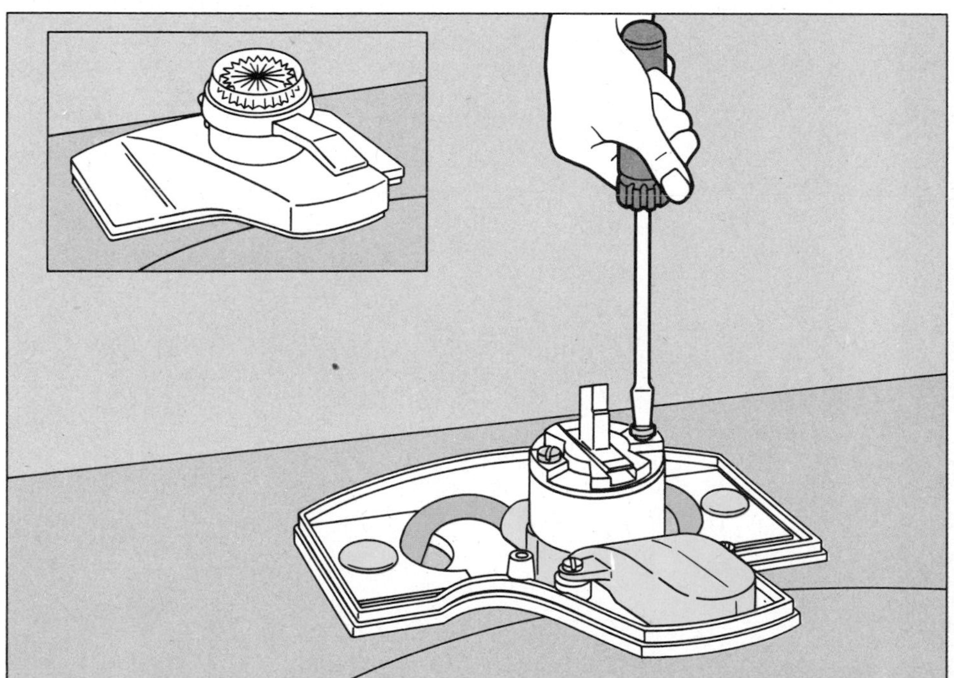

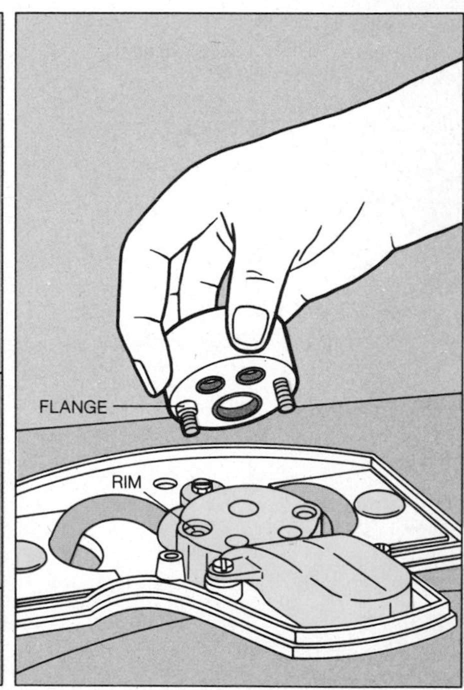

1 Removing the cartridge. Once you have removed the handle and chrome body cover *(page 37)*, it is a simple matter to remove the two brass bolts that hold the cartridge in place. But before deciding that a new cartridge is needed, check to be sure that the leak (which in this faucet shows up around the body or as a puddle under the sink) is not caused by a piece of dirt lodged between the ceramic disks. Turn the faucet on full and move the handle from side to side several times. If the leak persists, buy an identical cartridge for replacement.

2 Installing a new cartridge. Align the three ports on the bottom of the cartridge with the three holes in the faucet body. Make sure that the flange on one of the boltholes in the cartridge fits into the rim around one of the two boltholes in the faucet body. Replace the cartridge bolts, screw the Phillips-head screws up through the faucet body into the body cover, and tighten the setscrew in the faucet handle.

Some Special Twists for Tubs and Showers

The valves and spouts for tubs and showers are serviced like those on lavatories —with two exceptions. The diverter valve, which directs water to tub or shower, is special. And you may encounter an access problem repairing leaks from a wall-mounted stem valve.

To remove a wall-mounted faucet or diverter valve, you will have to unscrew a bonnet nut that is recessed. In some fixtures the out-of-the-way nut offers enough purchase to permit the use of locking-grip pliers or a basin wrench. If not, you will need a socket wrench to slip over the protruding faucet stem and the nut. Inexpensive sets of long sockets designed explicitly for this sort of job are available at plumbing-supply stores. Alternatively, your own tool kit may hold a socket that will suffice, although an unorthodox turning technique will be necessary because of the length of the faucet's stem *(right)*.

Unlike faucets and wall-mounted diverter valves, the working parts of a tub-spout diverter—a knob on top of the tub spout that controls a diverter gate within —are not replaceable. Failures can be remedied only by installing a new spout. Most problems with shower heads, however, are easily cured. Replacement of washers or O-rings will generally take care of leaks. And erratic or weak pressure from the shower head can usually be traced to a build-up of minerals from the water supply. Proper flow can be restored by disassembling the parts and giving them a good cleaning; soaking in vinegar loosens mineral deposits.

Wall-mounted Faucets

Removing the bonnet nut. The recessed bonnet nut of a wall-mounted fitting is generally plastered in place to keep splashing water from leaking into the wall. Chip away the adjacent plaster with a hammer and cold chisel so that the socket wrench—a type designed for plumbing jobs *(below)*—can be slipped over the faucet stem and bonnet nut. Insert the socket wrench's handle through the holes at the socket's outer end and turn to loosen the nut. After the repair has been made, replaster as necessary.

Shorter sockets that are normally used for work on engines can also be used to unscrew a recessed bonnet nut *(inset)*, since the faucet stem will fit through the hole meant for the tool's ratchet handle. In lieu of the handle, simply turn the socket with a pipe wrench.

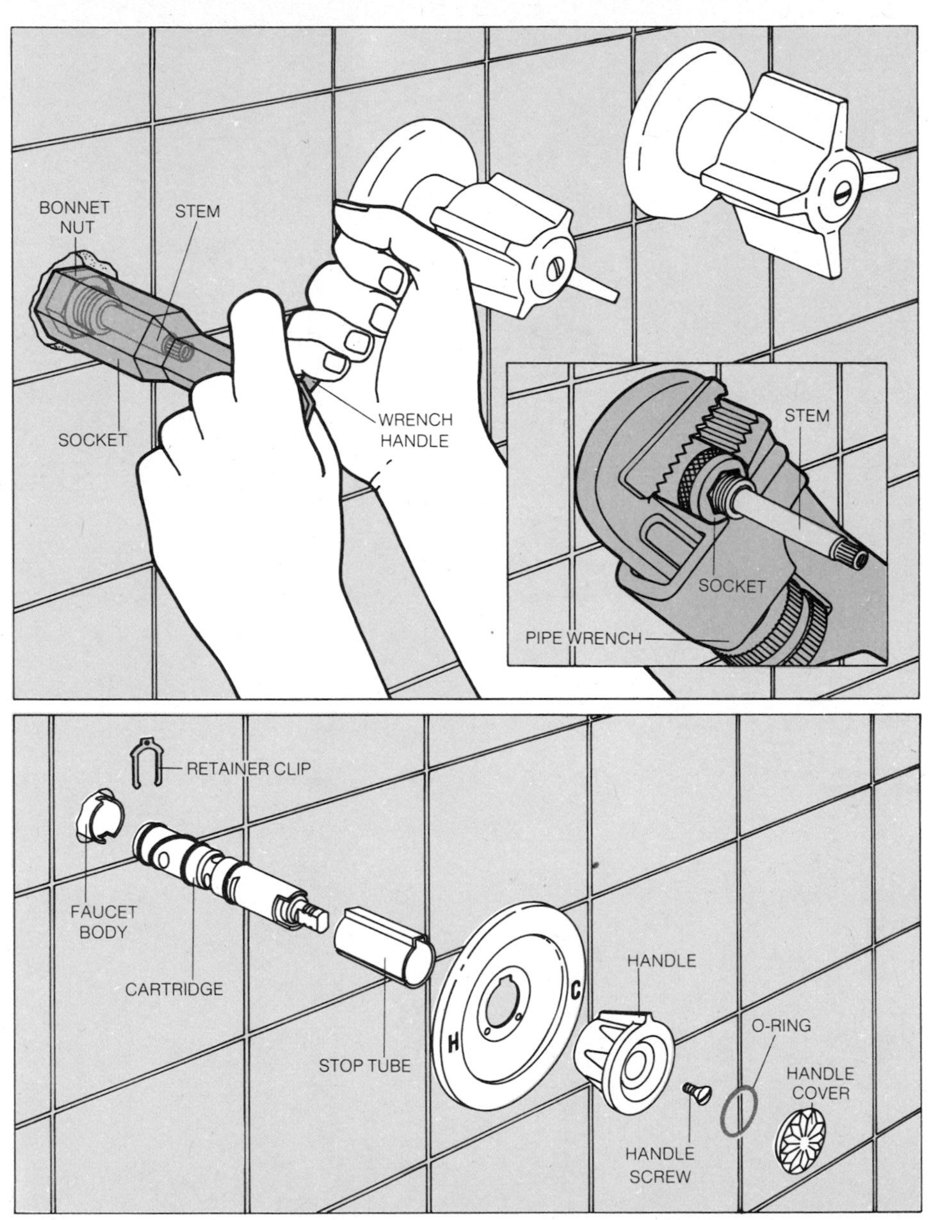

Fixing the faucet. The single-handle controls found on many modern tubs and showers work in exactly the same way—and are repaired the same way—as the single-lever faucets on sinks or lavatories *(pages 37-41)*, although the exterior portions may be quite different in appearance and assembly. To get at the mechanism of this particular faucet, pry off the handle cover with a small screwdriver. Unscrew the handle screw, remove the handle and escutcheon, then draw out the retainer clip with long-nose pliers, noting its position. Pull out the cartridge and replace it *(page 41)* with a new one, then reassemble the faucet. Stem-type tub and shower faucets are also repaired like their deck-mounted equivalents, as described on pages 30-35.

Tub-Shower Diverters

Repairing a diverter valve. A tub-shower diverter valve functions like a faucet. A clockwise turn moves the stem into the valve seat, closing the pipe to the tub spout and forcing the water instead through a hollow plastic housing and up into the shower head. A counterclockwise turn of the handle moves the stem back to open the pipe to the tub spout. If the valve leaks, disassemble it by the same steps used for the stem faucet on page 30. Replace worn washers, O-rings, packing, or badly worn metal parts. If the hollow housing has worn out so that the flow of water is incompletely diverted to the shower head, replace the whole diverter valve assembly.

Replacing a tub-spout diverter. The knob of a tub-spout diverter raises an internal gate that closes the pathway to the tub spout, forcing the water up to the shower head. If the mechanism fails to work properly, the entire spout must be replaced. To remove the old spout, place a piece of wood, such as a hammer handle, in the spout and turn it counterclockwise. Buy a replacement that is the same length as the old spout.

If a spout of the same size is unavailable, get a nipple—a short length of pipe that will make the connection to the pipes behind the wall *(inset)*. Use pipe-joint compound and lampwick to seal the nipple *(page 78)*. When screwing the spout into the nipple, hand-tighten it only. If alignment cannot be completed by hand, use the makeshift wood tool *(right)* or—provided the spout comes with a pad to protect the finish—use a wrench.

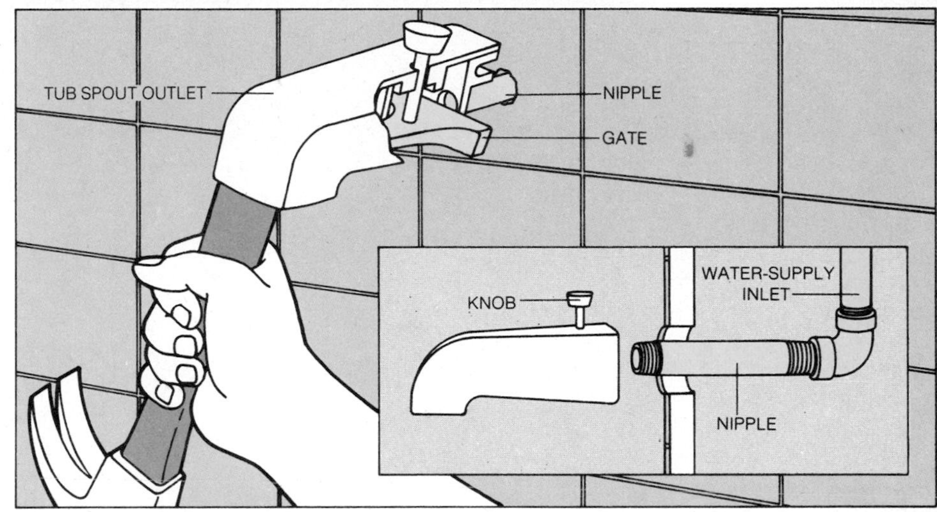

Shower Heads

Keeping a shower head showering. If dripping occurs where the shower head meets the arm, unscrew the head with tape-wrapped pliers. Replace the O-ring or washer that fits between the swivel ball and the outer part of the head; on the model shown you must first unscrew the collar.

If the flow of water is blocked, disassemble the shower head. Most have a screw that secures a face plate or a screen cover. The adjustable type in the drawing also has a handle operating a cam that moves adjusters in and out of the face-plate holes to make the spray finer or coarser. As you remove parts, line them up in order for easy reassembly. To clean off mineral deposits, soak the parts in vinegar. Use a toothpick to clear the holes of the spray adjusters or screen, and scrub the other parts with a wire brush. Before reassembling, lubricate the swivel ball with petroleum jelly. When screwing the shower head back into the arm, hand-tighten it only, using pipe-joint compound to seal the joint *(page 78)*.

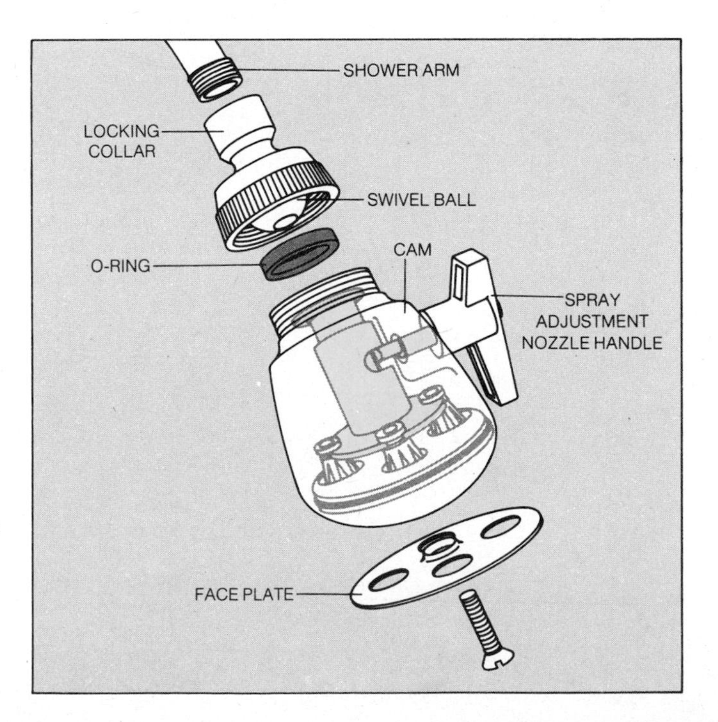

Beyond the Faucet: Spouts, Stoppers and Drains

Water released by a faucet makes a circuitous journey. It pours out of a spout, shower head or kitchen-sink hose, sometimes going through an aerator where it mixes with air to produce a splashless stream, then flows past a stopper or pop-up plug, onward through a strainer at the base of the bowl or tub into a water-filled trap below, and finally enters the house drainage system.

Compared to the faucets that set the whole process going, these flow-and-drain fixtures are relatively simple. Locating a trouble spot is easy and the jobs that must be done are usually straightforward repairs or replacements. You do not have to shut off the water supply for these jobs—just close the faucets tight.

The difficulty of the jobs, when difficulty arises, is in getting at a fixture and reassembling its components in the correct order. Some of the fixtures are nestled under sinks, basins and tubs, where work space is cramped and special tools may be needed to unscrew fasteners. Others consist of intricate combinations of small parts, which must be fitted together precisely.

The combination of sink spout, aerator and spray shown here presents the full range of these problems. An aerator unscrews easily from the end of a spout and should be removed periodically for cleaning, because minute amounts of grit in the water supply will quickly clog it. But an aerator will not do its job if its internal parts are replaced incorrectly.

The spray head also contains an aerator; here, clogging can block the action of the diverter valve that switches water from spout to spray. Concealed in the base of the spout, this valve is the most delicate component of the entire assembly. Like aerators—though far less often—it can clog up, and even a clean valve will not work if its covering fills with grit or dirt. If you have cleaned both aerators and still have problems—low or uneven water pressure, or a failure to switch smoothly from spout to spray and back again—go to work on the diverter valve.

The sturdiest, simplest component of all—the spray hose—is, paradoxically, the hardest to work with. Replacing the hose calls for tight maneuvering under the sink, often between two adjoining faucet pipes. In these close quarters the plumbing tool called a basin wrench may offer the only way of getting at the nut that holds the hose in place.

Cleaning a Spray Nozzle

An array of small parts. Unscrew the aerator from the end of the spout with tape-wrapped pliers. Disassemble the parts inside the aerator body and set them aside in the correct order and orientation: a part reassembled upside down will keep the aerator from working. Whatever its design, your aerator will contain a plastic or rubber washer, a disk perforated by tiny holes or a sawtooth edge, and one or more screens. A more complex model *(insert)* may also have a disposable adapter for internal and external spout threads, and air-intake holes in the outer shell. Clean the screens with a small stiff brush; use the brush and a toothpick to clean out disk and intake holes. Replace badly worn or misshapen washers, and flush out all parts by holding them upside down in a full stream of water before reassembling the aerator.

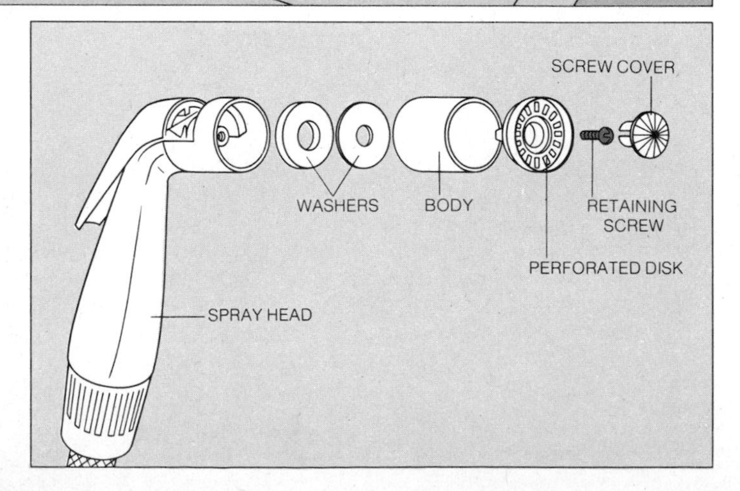

A simplified aerator. On some models you can unscrew the tip of a spray nozzle by hand. More often it is secured by a Phillips screw and the screw may be concealed by a cover. Pop the cover out with a screwdriver or penknife, remove the retaining screw and disassemble the internal parts. Clean or replace these parts as you would those of an aerator; before reassembling them, run water through the spray head at full force for a minute or two. Do not try to repair the mechanism inside the spray head. If defective, this sealed unit should be replaced with an identical model from the manufacturer.

Removing the Hose

1 At the spray head. If a hose leaks or blocks water, detach it for possible replacement, starting at the spray head. Unscrew the head from its coupling, then free the coupling from the hose by prying off a retaining snap ring with the tip of a screwdriver or penknife. Replace the hose washer if necessary, and try to clear a blocked hose by running water through it at full force (with the spray head removed, opening a faucet will send water to the hose rather than the sink spout). If the hose still leaks or is permanently obstructed, go on to Step 2.

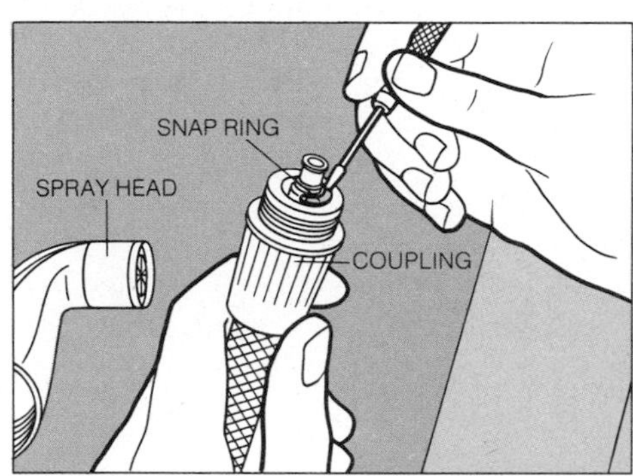

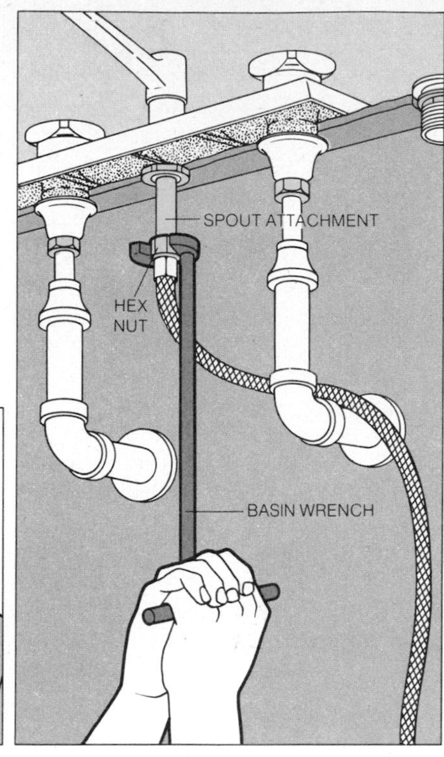

2 At the spout. Only a single hex nut secures the hose to its attachment at the base of the spout, but this nut is often hard to get at. If you cannot unscrew it with an ordinary wrench or a pair of locking pliers, use a basin wrench *(page 10)*, a plumber's tool especially designed for working in close quarters. Lie on your back under the sink as you unscrew the nut, and illuminate the work area with a work light or a flashlight.

Replace the hose with a new one, preferably of plastic vinyl with nylon cord reinforcement. Be sure to take the hex nut with you when you get the new hose; if your plumbing supplier does not have a model that matches your spout attachment, he will furnish you with an adapter.

Working on the Diverter Valve

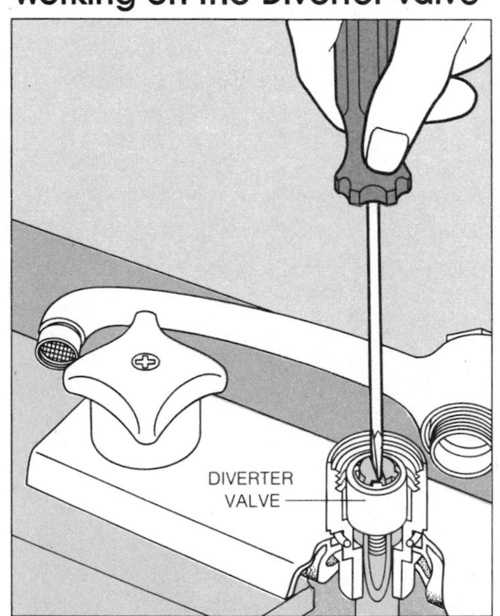

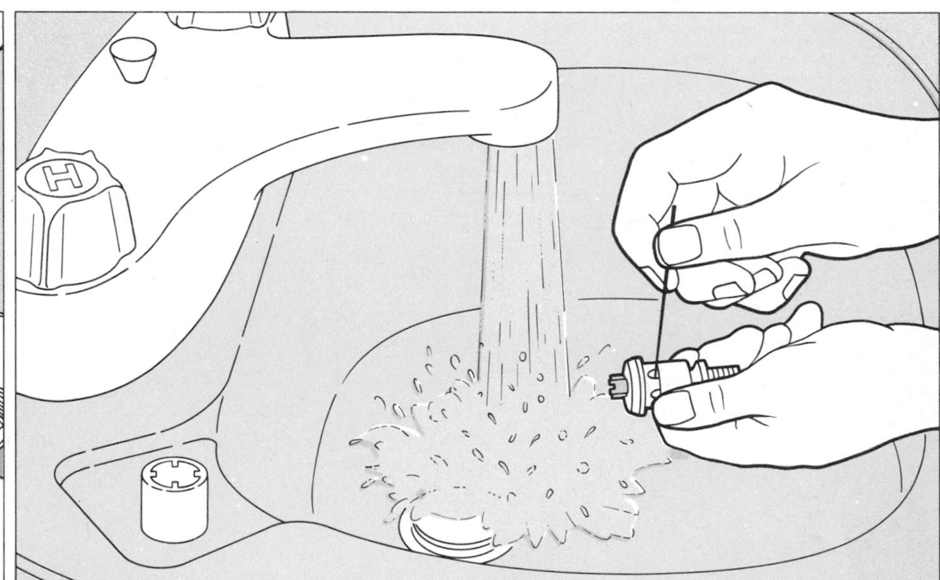

1 Getting into the spout. In some sinks the spout is secured to the faucet body by a grooved ring, in others by a nut atop the spout. These chrome-plated fasteners are easily scarred; remove them carefully with a tape-wrapped wrench or pliers. Inside the faucet body you can now see the tip of the diverter valve, usually capped by a brass screw. The screw is an integral part of the valve: turn it just enough to free the valve from the valve seat inside the faucet body, then pull out the screw and valve together. A valve without a screw top can simply be pulled straight out with pliers.

2 Cleaning the valve. The valve will have a distinctive pattern of small outlets and channels and may come as a removable inner body sliding in a sleeve. Take it apart, if possible, and clean all its openings and surfaces with a sharp, soft object, such as a toothpick. Do not use a metal tool for this job; the valve is easily nicked or scratched. As you work, flush the valve frequently with water at full force from a working faucet. Reassemble the valve and spout, and try the spout-spray mechanism. If switching is poor or pressure uneven—and if you have already checked out the aerators, the spray head and the hose—replace the diverter valve. Take the old valve with you to your plumbing supplier to be sure of getting a perfect match.

Replacing Sink Strainers

If you notice a leak around a sink drain hole, you may be able to fix it by simply loosening the strainer from underneath and applying fresh plumber's putty under the lip. Often, however, the strainer itself is corroded—it may break apart as you pry it up. Replacement is simple and inexpensive. There are two kinds of basket strainers: one secured by a lock nut, and the other held by a plastic retainer and three screws. The lock-nut type is generally used for stainless-steel sinks because the lock nut tightens against the sink without bending the sink metal.

To remove an old strainer, first detach the tailpiece *(page 51)*. Then remove the lock nut or detach the retainer screws and pry the old strainer out.

Installing a Lock-Nut Strainer

1 Sealing the opening. Turn off the water supply. Remove all old putty from around the drain opening in the sink and dry it completely. Apply a ⅛-inch bead of plumber's putty to the flange of the opening and place the strainer body through the opening, pressing down firmly so that the putty spreads evenly.

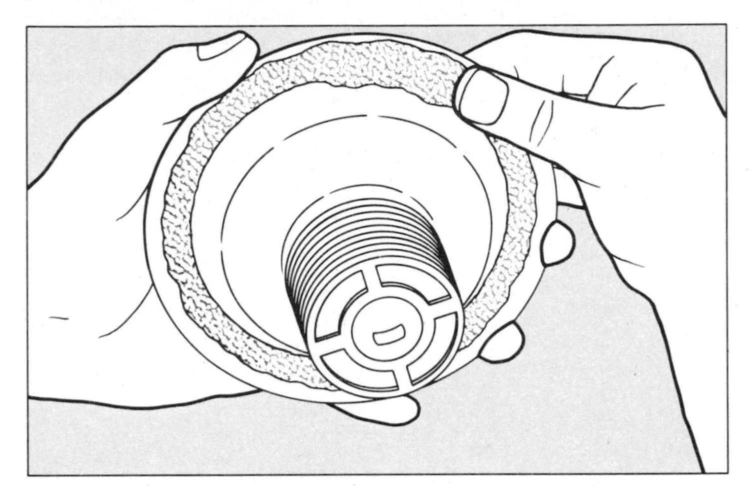

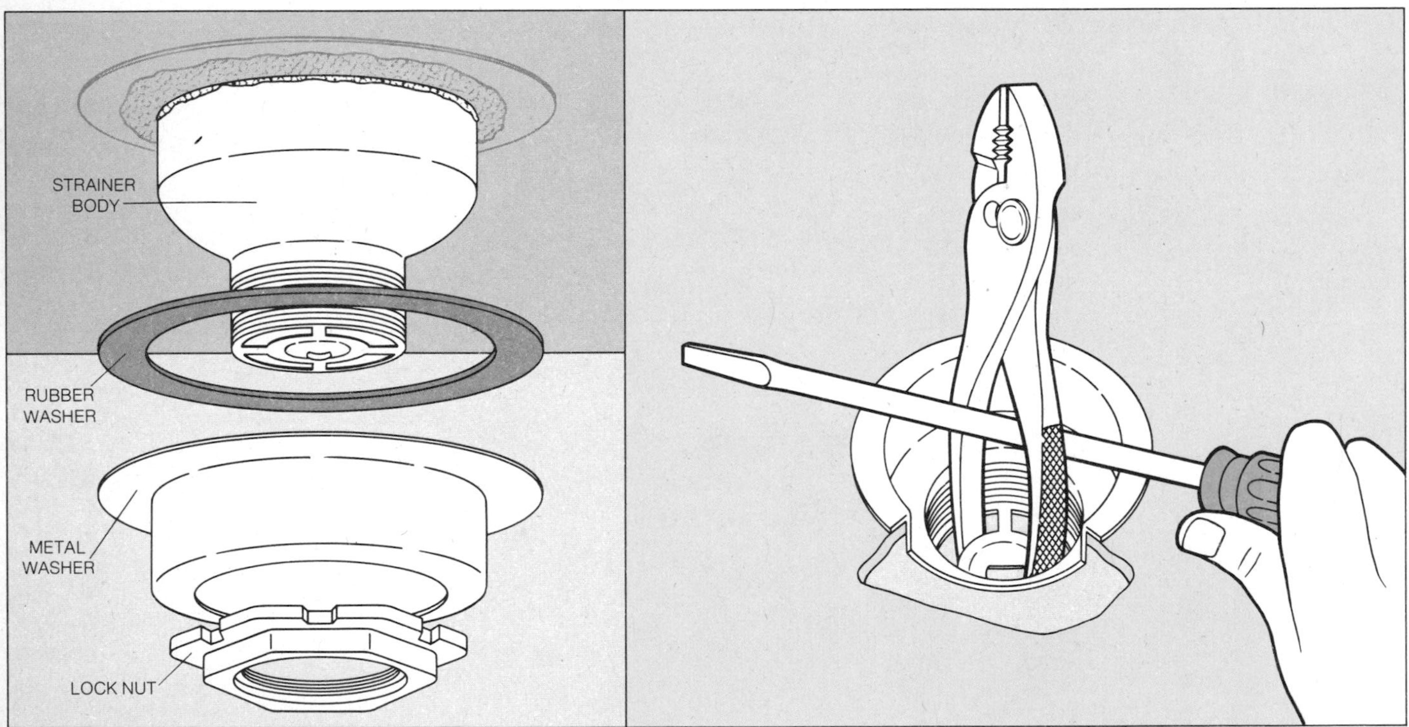

2 Securing the strainer. Place the rubber washer and the metal washer onto the strainer body. Then screw on the lock nut finger-tight *(above)* to hold the strainer while you work underneath the sink. Place the handles of pliers into the crosspieces of the strainer and slide a screwdriver between the handles *(right)*. Hold onto the screwdriver with one hand to immobilize the strainer while you tighten the lock nut.

3 Tightening the lock nut. Tighten the strainer several turns more, using a 14-inch pipe wrench or a hammer and wood dowel as shown below. Brace the dowel against one of the grooves of the lock nut and tap it with the hammer. Do not overtighten the lock nut because you may distort the metal parts or crack the ceramic.

4 Connecting to the trap and tailpiece. If the tailpiece is worn or corroded, replace it *(page 51)*. Otherwise, fit the strainer sleeve over the existing one, and secure it by tightening the coupling. Then tighten the trap coupling. Wipe away excess putty with a soft cloth. Turn on the water and check to be sure there are no leakages.

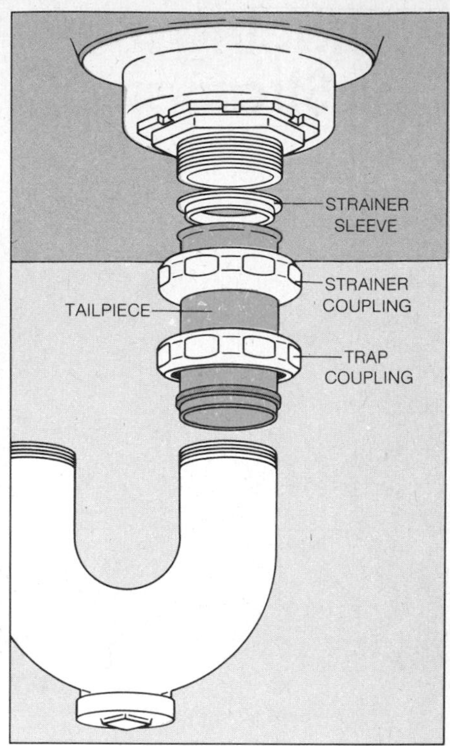

Installing a Retainer-type Strainer

Attaching the retainer. Put the strainer body into the puttied opening, then attach the rubber and metal washers from underneath as for the lock-nut strainer *(opposite, bottom)*. Fit the retainer onto the strainer body and turn it until the ridges on the side of the drain fit into the grooves of the retainer. Twist to lock it in place, then tighten the retainer screws. Connect the tailpiece as you would for a lock-nut strainer.

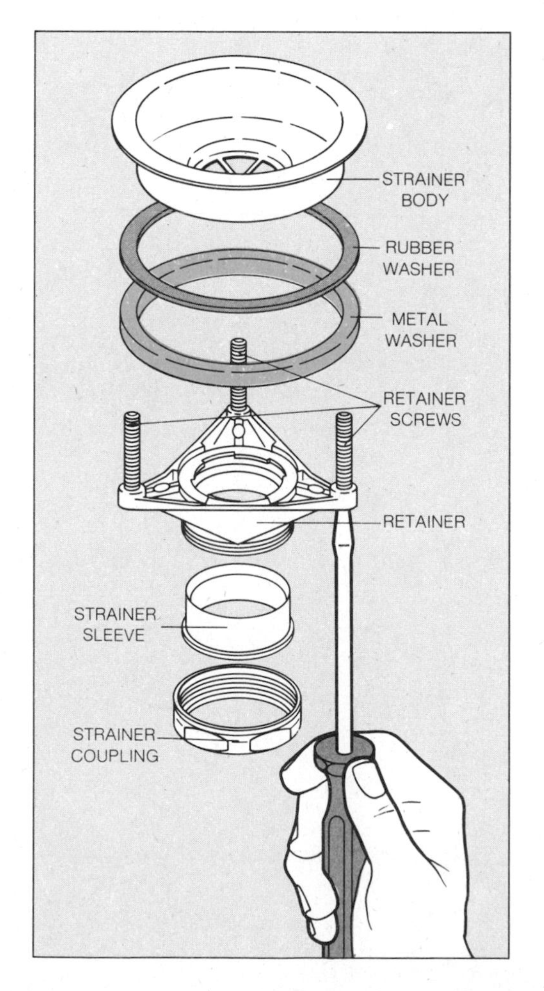

Adjusting and Replacing a Lavatory Pop-up

The seal of a pop-up drain—a metal plug closing upon a metal flange, or ring—is never quite as watertight as that of an old-style rubber stopper. Moreover, the pop-up mechanism has several moving parts and needs periodic adjustments. But the pop-up's convenience has made it almost universal, and a homeowner must learn to cope with its malfunctions. They are, fortunately, few in number: a plug that fails to open or close properly, and the two enemies of all drainage systems—clogs and leaks.

Pop-up problems are usually caused by faulty connections. The control knob atop the lavatory is part of a three-section linkage: a vertical lift rod; another vertical rod, flat and pierced by holes, called a clevis; and a seesaw-like horizontal rod that pivots on a plastic ball inside the drain to raise and lower the plug. Adjusting the mechanism calls for two simple settings on the lift rod and clevis.

Loose hair and similar lavatory debris are the usual causes of clogging. To clear the drain you must remove the plug, either by lifting or twisting it out of the lavatory bowl, or by disassembling the pivot rod *(opposite, top)*. When you have freed the plug, clean it thoroughly, then clear out the drain below with a brush or piece of cloth wrapped around stiff wire, such as a length of coat hanger.

There are two kinds of leaks. Water that drips or trickles from the mechanism beneath the bowl is leaking around the pivot ball. Tighten the retaining nut that holds the ball in place; if that does not work, remove the nut and then replace the pivot-ball washer. Water that seeps down the outside of the drain is a more serious matter. The thudding of the plug against the flange may have broken a putty seal beneath the flange; by loosening a lock nut under the bowl you can lift the flange slightly and renew the seal. More often, this type of leak is due to corrosion of one of the parts of the drain—a sure signal that the drain will soon fail completely. To fix the leak properly, you should replace the entire drain assembly *(opposite, bottom)*.

1 Setting the lift rod. If the pop-up plug does not make a good seal in the closed position, begin your adjustment with the lift rod—the vertical shaft that runs down from the control knob and through the top of the lavatory. Pull the knob as far up as you can, and free the rod by loosening the clevis screw with a pair of tape-wrapped pliers. Press the pop-up plug down to seal the drain (the clevis rod will rise a bit) and retighten the clevis screw. You may find that the pop-up mechanism is now slightly jammed and difficult to operate; if so, go on to Step 2.

2 Setting the pivot rod. With your fingers, squeeze the spring clip that holds the pivot rod in the clevis rod completely out of the clevis. Reset the rod in the next higher clevis hole, threading it through the spring clip on both sides of the clevis. Try the pop-up mechanism and, if necessary, repeat Step 1.

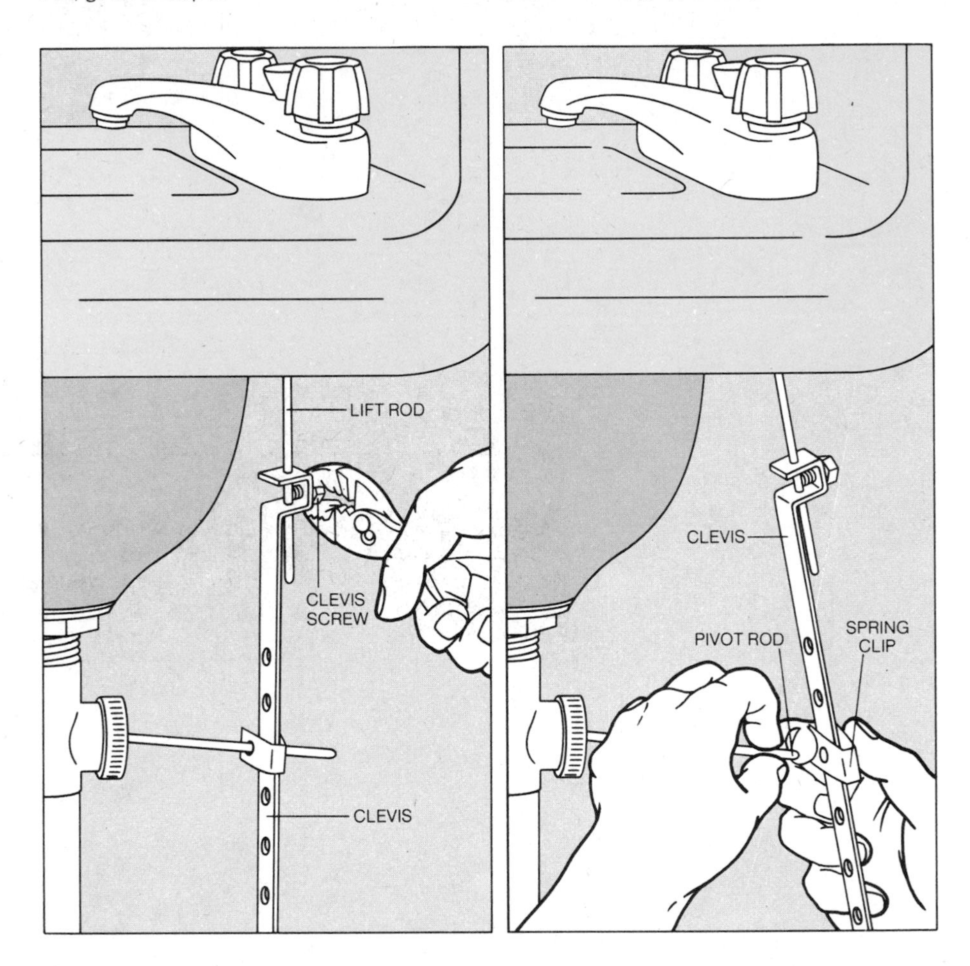

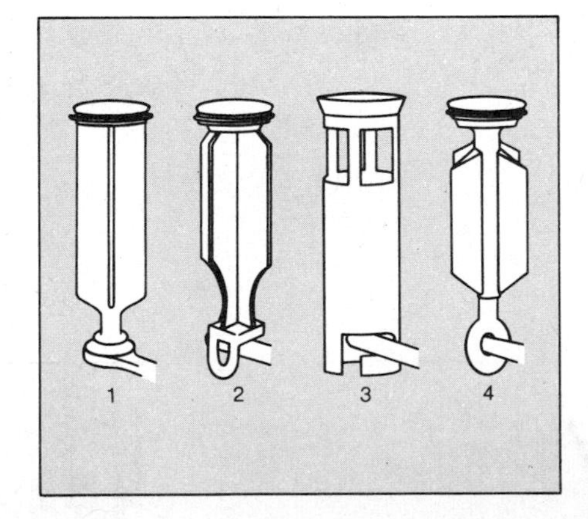

A medley of pop-up plugs. These four common types of pop-up plugs all do the same job, but the methods of installing and removing them differ widely. Numbers 1 and 2 rest on the inside end of the pivot rod; to remove either of them, raise the plug to the open position and lift it out of the drain. Number 3 engages the rod in a slot, and comes free of the rod with a quarter turn counterclockwise. If you cannot remove your pop-up plug by lifting or turning, it must resemble Number 4, which engages the rod in a loop like the eye of a huge needle. To disengage this plug, you must free the pivot rod and pull it partly or completely out of the drain T underneath the basin *(opposite, top)*.

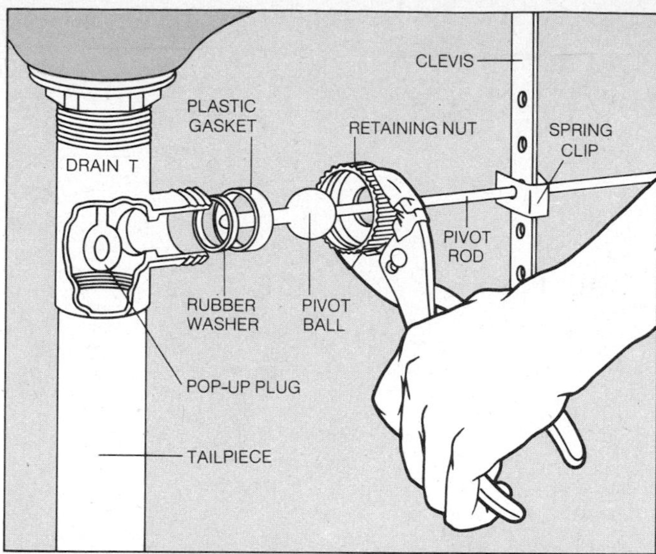

The pivot-rod assembly. With a tape-wrapped adjustable wrench or pliers, unscrew the retaining nut that secures the pivot-rod assembly inside the drain T. Squeeze the spring clip on the clevis and back the pivot rod out of the T. You can now pull the plug out of the drain for cleaning.

Follow the same procedure as in the first step in dismantling a pop-up drain, either to stop leaks or to install a new assembly. A leak at the retaining nut may be stopped simply by tightening the nut, but check the gasket and washer inside the pivot ball and, if necessary, replace them. The remainder of the disassembly and installation procedure is described below.

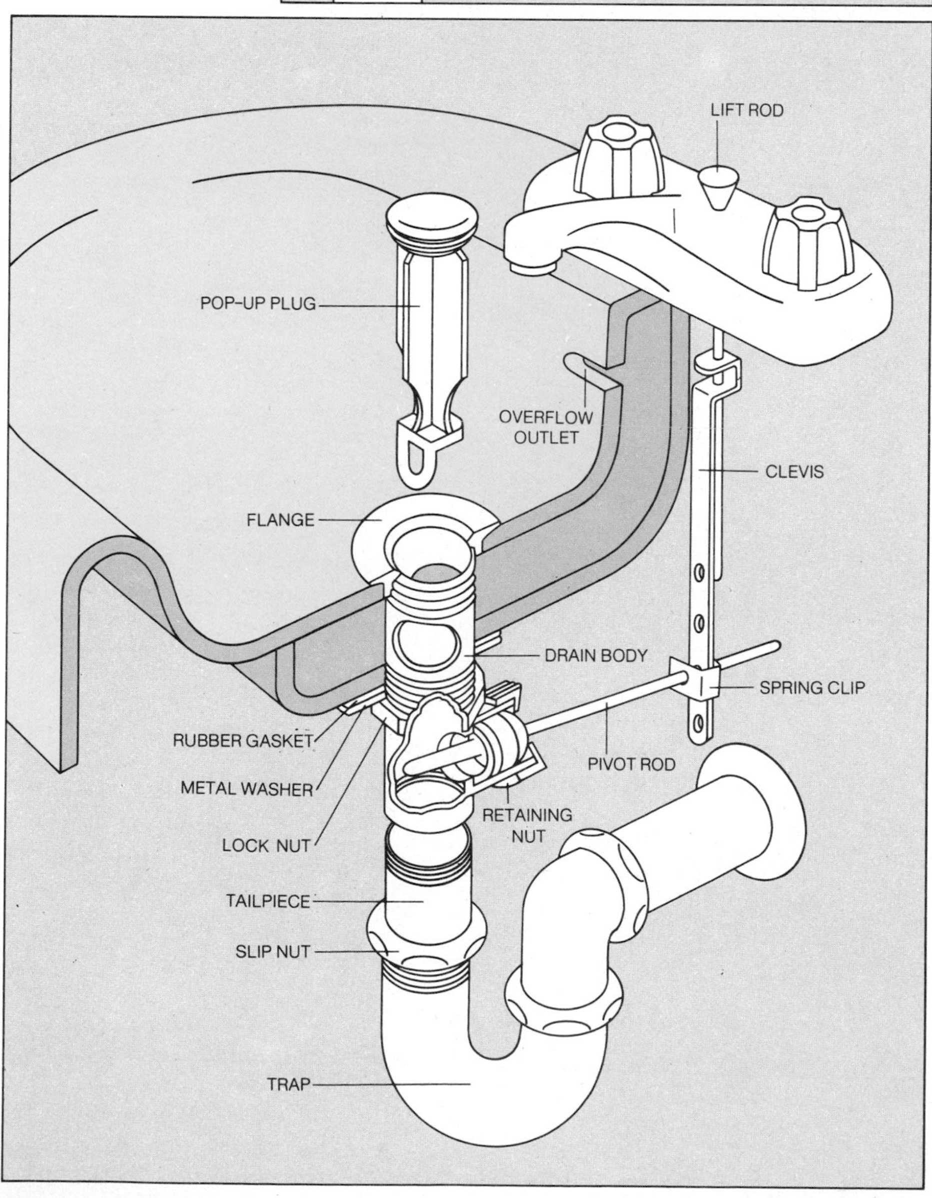

Replacing the drain. Start a replacement job by freeing the pivot rod from the drain T *(above)*. Unscrew the slip nut that fastens the tailpiece to the trap, then unscrew the tailpiece from the drain body and push it down into the trap, out of the way. Loosen the lock nut that secures the drain body underneath the basin, and unscrew the drain body from the flange. An old, corroded body will be hard to budge at first; do not hesitate to apply pressure with the wrench, even if you flatten the body a bit. Inside the sink pry up the flange with a dull-bladed knife or the tip of a screwdriver; be careful not to scar the sink surface. Scrape away the putty at the mouth of the drain and wipe the drain clean with a dry cloth. Discard the old tailpiece (a new one will come with the new drain assembly) but keep the slip nut that fastened it to the trap; replace the slip-nut washer if it is badly worn.

Use tape-wrapped wrenches for all steps of the installation job. Start by setting a slip nut and a washer on the new tailpiece and pushing the piece down into the trap. Apply a ⅛-inch bead of plumber's putty under the rim of the flange, and press the flange into place in the mouth of the drain. Coat the threads of the drain body and tailpiece with pipe-joint compound, screw the drain body into the flange and the tailpiece into the drain body, and tighten the tailpiece slip nut onto the trap. Finally, tighten the lock nut against the washer and gasket with an adjustable wrench. Caution: Do not overtighten this nut or the porcelain above it may crack.

If you are replacing the pop-up mechanism as well as the drain, feed the pivot rod into the drain T and tighten the retaining nut. Insert the lift rod down through the faucet body or the top of the basin and fasten its lower end to the clevis with the clevis screw. Feed the pivot rod through a clevis hole, making it fast with the spring clip. Try the mechanism and, if necessary, reset the lift rod and clevis *(opposite, top)*.

Replacing Traps and Tailpieces

The curved traps beneath kitchen sinks and lavatories are vital but vulnerable. Some of the water that flows off through the sink or lavatory drain remains in the trap as a seal to keep odors and gases from backing into the house from drain and sewer lines. But traps are often made with comparatively thin walls because they are not subjected to the high water pressure of supply lines. In time, they corrode and leak, and must be replaced.

The traps shown on these pages, called P traps, are the ones commonly found in most homes. They may be made of brass, galvanized iron, steel or (where plumbing codes permit) plastic. The best—and most expensive—are chrome plated for looks and relatively heavy for long wear. Choose the heaviest traps you can afford; they are a better buy, particularly on fixtures that get substantial use. Choose a matching material and weight for the pipes that accompany a trap: the tailpiece that connects the trap to the sink or lavatory drain, and the drainpipe that connects the trap to a drain outlet.

The trap you choose may be either a swivel or fixed type. A swivel trap can be turned in any direction on a drainpipe by adjusting a separate slip nut—a useful feature when you want to replace a tailpiece without first removing the trap, or when you must make a connection between a drainpipe and sink that are not in perfect alignment.

Fixed traps, which screw directly onto the drainpipe, are less adaptable. You can, however, replace a tailpiece on a fixed trap without removing the trap by using a professional plumber's tactic *(opposite, bottom right)*. Plumbers "roll the trap"—that is, they free the tailpiece and drop it neatly into the trap; then loosen the trap and roll it to one side to get at the tailpiece. The procedure may seem tricky, but it is surprisingly easy to follow, and it makes a tailpiece replacement as simple on a fixed trap as on a swivel. Replacing drainpipes, on the other hand, remains a tougher job: no matter what type of trap you have, you must remove it completely to get at the drainpipe.

Three Trap-to-Drain Connections

Removing a one-piece swivel trap. Set a pail beneath the trap and remove the cleanout plug, if there is one, to drain the trap. With a tape-wrapped monkey wrench or channeled pliers, unscrew the nuts that attach the trap to the tailpiece and drainpipe. If the drainpipe must be replaced, pull the escutcheon from the wall and unscrew the slip nut that fastens the drainpipe to the drain outlet. Discard all the nuts and washers and clean the drain outlet threads.

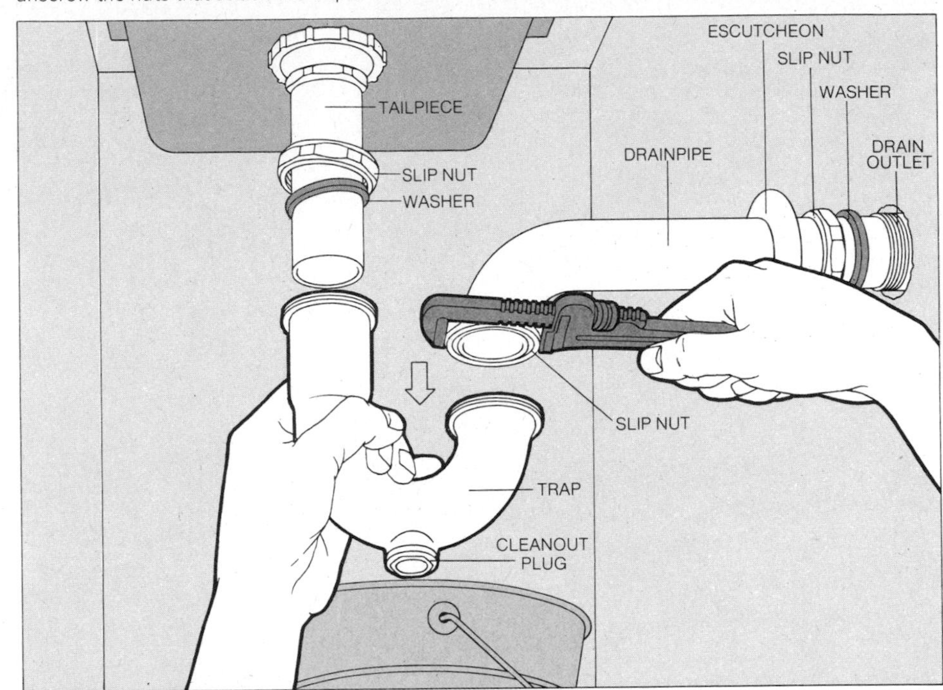

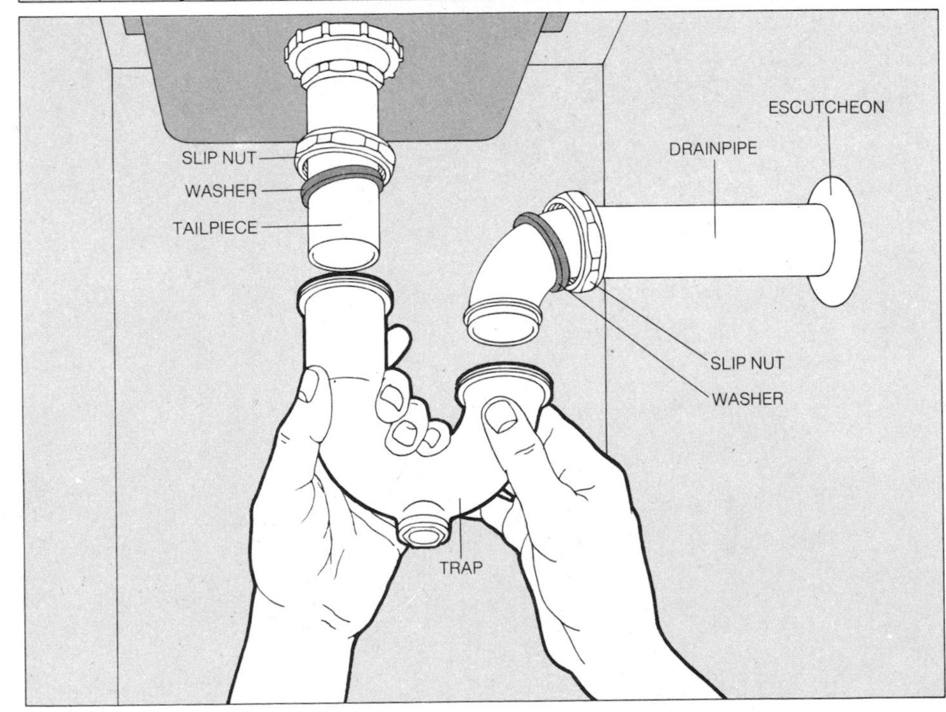

Installing a new trap. If you are replacing the drainpipe, slide the escutcheon, a slip nut and a washer—in that order—onto the end of the new pipe that will face the wall. Screw the slip nut tight to the drain outlet and push the escutcheon against the wall. Slide the slip nuts and washers that come with the new trap onto the tailpiece and the drainpipe, slip nuts first.
Fit the trap in place against the tailpiece and drainpipe, and tighten the slip nuts.

A two-piece swivel trap. Remove the U-shaped section of a two-piece trap by unscrewing the slip nuts at the ends, and sliding them back onto the tailpiece and the trap. The elbow screws directly onto the drainpipe; remove it by turning it counterclockwise. To install a replacement, first separate the elbow from the U-shaped section. Coat the threads of the drainpipe with pipe-joint compound or plastic joint tape *(page 78)* and screw the elbow onto it. Slide a new slip nut and a washer onto the tailpiece, set the trap in place against the tailpiece and the elbow, and tighten the trap and tailpiece slip nuts.

A fixed trap. A slip nut fastens a fixed trap to the tailpiece; at its other end the trap screws directly onto the drainpipe. Before removing the trap, you must disengage the tailpiece running between the trap and the sink or lavatory drain. Unscrew the slip nuts that fasten the tailpiece to the drain and trap, push the tailpiece into the trap, then turn the entire trap counterclockwise with your hands or a wrench until it comes off the drainpipe. Transfer the old tailpiece to your new trap and coat the drainpipe threads with pipe-joint compound or plastic joint tape before making a replacement.

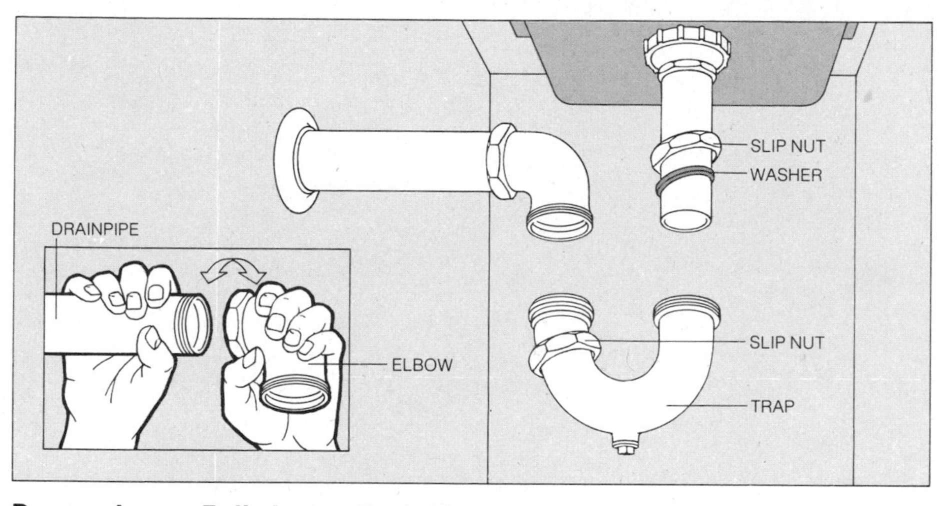

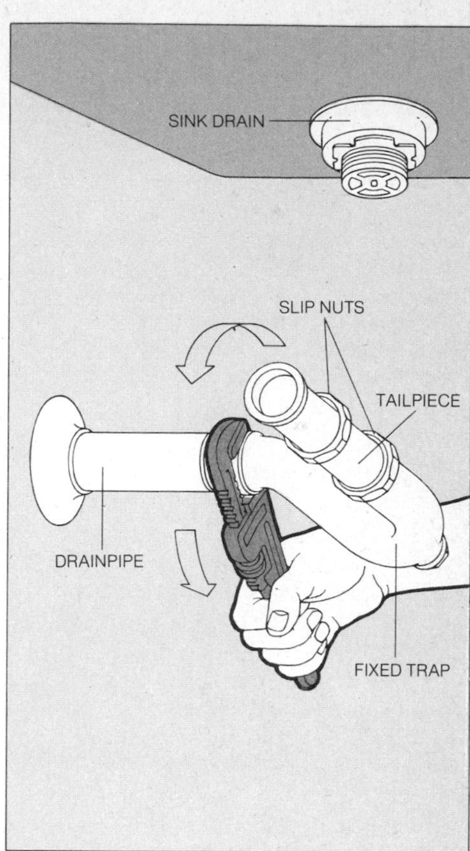

Removing a Tailpiece: Two Shortcuts

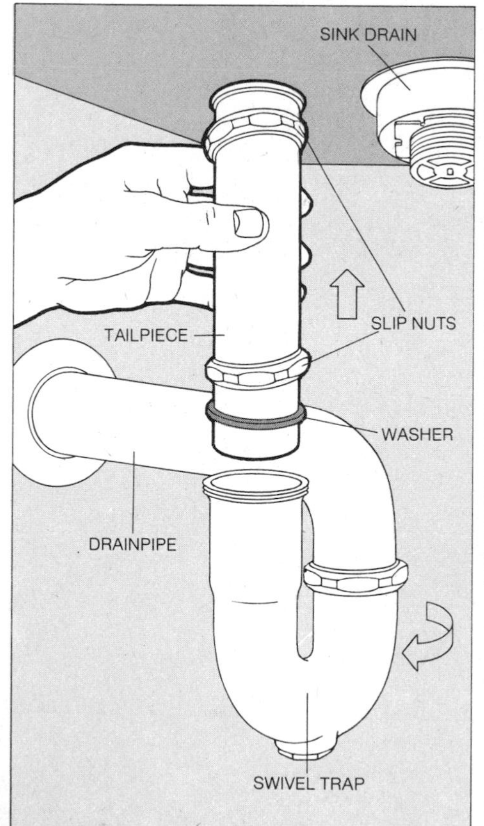

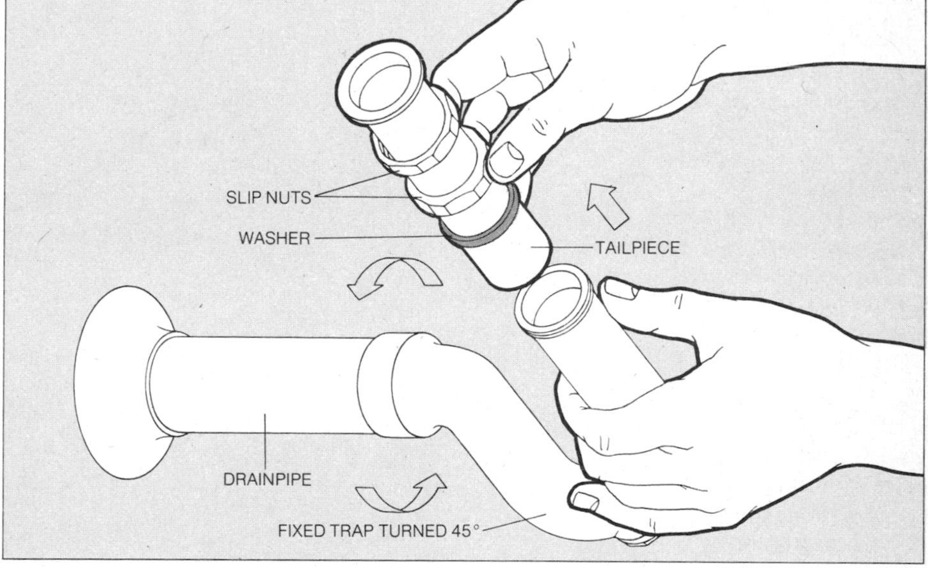

On a swivel trap. Unscrew the slip nuts that fasten the tailpiece to the trap and the sink drain, and push the tailpiece down into the trap. Loosen the slip nut that fastens the trap to the drainpipe or an elbow, and swivel the trap away from the sink drain, then lift out the old tailpiece and insert the new one. To complete the installation, lift the new tailpiece up to the sink drain and tighten all slip nuts.

On a fixed trap. Free the tailpiece as you would on a swivel trap *(left)* and push it down into the fixed trap. Then, using your hands or a tape-wrapped wrench, turn the entire trap counterclockwise about 45° on the drainpipe. You can now pull the old tailpiece out of the trap and insert the new one. Complete the installation by retightening the trap on the drainpipe and fastening the slip nuts on the tailpiece.

Adjusting Bathtub Drains

Modern bathtub drains, controlled by a lever on the overflow plate, operate in large part from a position of concealment. Hidden in the bathtub overflow tube is a so-called lift linkage that, rising or falling in response to the control lever, opens or closes the drain in one of two ways. A pop-up drain utilizes a metal stopper at the tub outlet, while a trip-lever drain regulates the outflow of water with a plunger at the intersection of the overflow tube and the drain.

A common problem of bathtub drains is clogging caused by the accumulation of hair on the trip-lever drain plunger or on the spring at the end of a pop-up drain lift linkage. To remove the hair, the lift linkage must be removed from the overflow tube. This is done by unscrewing the overflow plate and pulling on it. The same procedure is sometimes necessary in order to adjust the length of the lift linkage; improper adjustment—perhaps caused by faulty installation, perhaps by wear and tear—can result in a leaky drain or one that fails to open fully.

The various conduits of the drain—the overflow tube, outlet pipe and trap—are less likely to suffer problems than those of lavatories or sinks. They are usually made of heavier pipe with sturdy cast-brass fittings that strongly resist corrosion. And their hidden position protects them from accidental knocks.

A trip-lever drain. The key element of a trip-lever drain is the brass plunger suspended from the lift linkage. When lowered by the control lever, the plunger sits on a slight ridge below the juncture of the overflow tube and the drain, blocking outflow of water via the main tube outlet. However, any water that spills into the overflow tube can pass freely down the drain because the plunger is hollow. If a drain of this type leaks, the cause may be wear on the plunger due to repeated impact against its rigid seat. Lift out the whole mechanism, and slightly lengthen the linkage, as shown opposite at top. At the same time, check the cotter pins of the assembly; if they are corroded, they should be replaced.

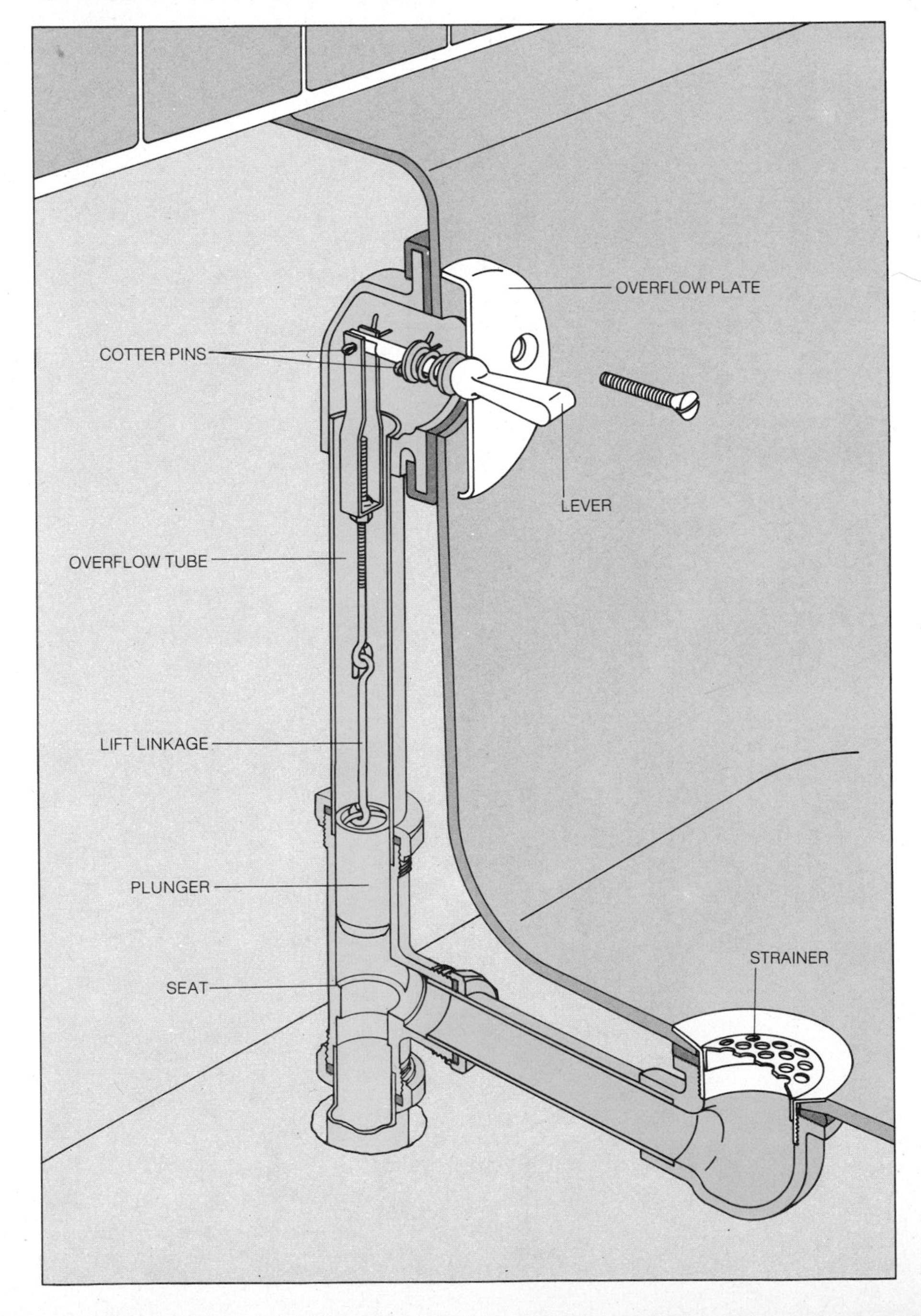

Adjusting the lift linkage. The upper segment of the lift linkage consists of a brass yoke from which a threaded rod is suspended; a lock nut secures the threaded rod in place. To adjust the length of the linkage, loosen the lock nut with pliers, turn the threaded rod the desired amount, then tighten the lock nut again. Try slight adjustments at first. A trip-lever lift linkage that has been lengthened too much will work fine when the drain is closed, but will fail to lift the plunger clear of the drain when the control lever is in the open position. Similarly, an excessively long lift linkage in a pop-up drain will prevent the stopper from completely shutting.

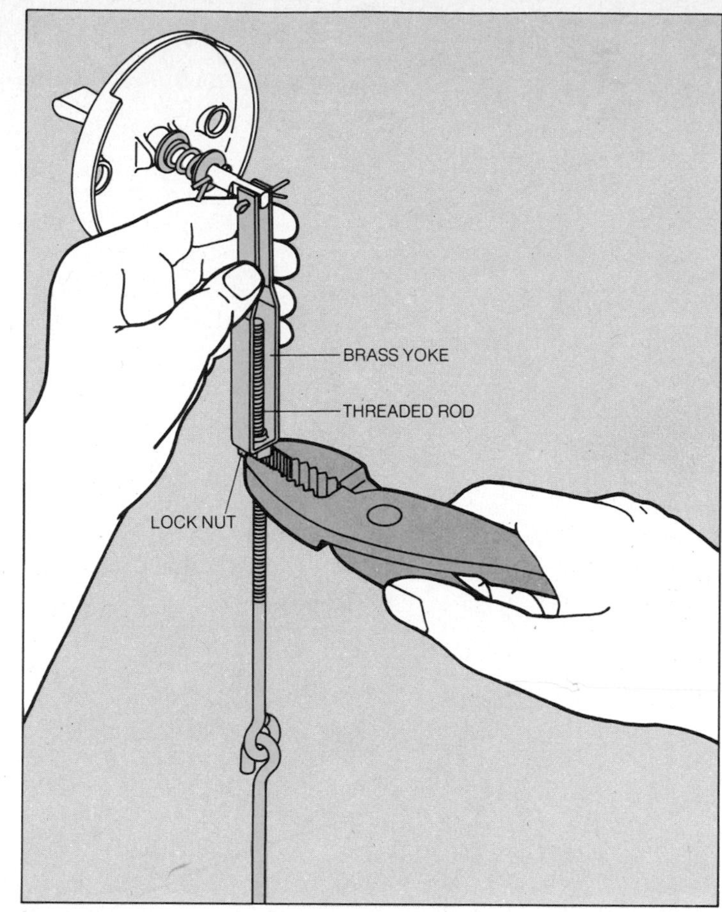

A pop-up drain. The lift linkage of a pop-up drain resembles that of a trip-lever drain, except that the lower end of the linkage is shaped to form a stiff spiral spring. This spring rests on the end of a separate horizontal linkage shaped like a rocker in the middle and leading to the metal stopper. When the spring presses downward, the stopper rises, seesaw fashion.

Because the pop-up drain lacks a screen at the tub outlet, the stopper has a cross-shaped base that prevents small objects—such as the top of a shampoo bottle—from passing down the drain. Householders sometimes try to cure sluggish drainage during a shower by entirely removing the plug, but this merely invites a more serious blockage. The proper solution to a persistent clogging problem is described on page 24.

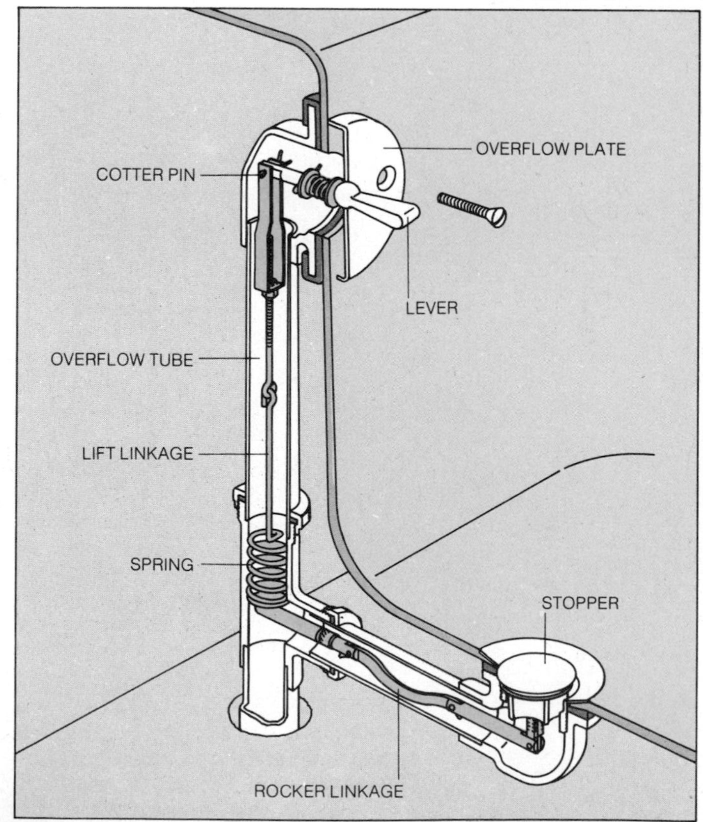

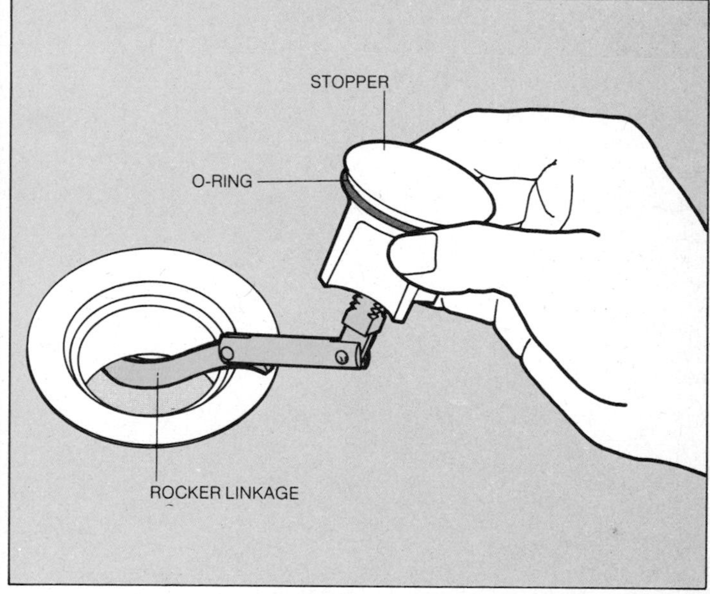

Pop-up repairs. If a pop-up drain leaks, the O-ring below the metal stopper may be worn. Open the drain and pull out the stopper and the rocker linkage. Clean these parts of accumulated hair. Slip on a new O-ring, then replace the stopper, working it sideways or back and forth until it clears the bend in the pipe. Make sure that the bottom of the curve in the linkage faces down.

Toilets: Simple Repairs for Complex Machines

Considering that nearly half the water that is used in the average household flows through toilets, it is remarkable that these virtually automatic devices give such trouble-free service with so little attention. Repairs and adjustments are usually minor and within the capabilities of an inexperienced home plumber.

Although the mechanism of the tank toilet, by far the most common type in houses, remains a mystery to many, no householder need be finicky about removing the tank top and poking around inside. The water there is as pure as it is anywhere in the house, and the valves, levers and floats move slowly so that it is easy to observe the way they interact and control the flushing cycle.

Leaks and noises are the most frequent problems with tank toilets, and the two are often interrelated. An intermittent gurgle of water from tank to bowl, for example, indicates a faulty outlet valve *(below, top)*. A high whine or whistle accompanied by a continuous run of water is a sign that the ball cock—the device that starts and stops the refill cycle—needs attention *(pages 56-57)*. Learning to diagnose such problems is a big step toward correcting them.

Another kind of toilet, the pressure flush valve type *(pages 61-63)*, has no tank but uses a pressurized flow of water to achieve the flushing action. Although rare in houses, pressure valves are common in apartments and other multiple-family dwellings because some models use less water than tank toilets. For the same reason, they have in recent years been recommended and sometimes required for new buildings in areas with chronic water shortages. While the valves are somewhat intricate in design, they are on the whole easier to adjust and maintain than tank mechanisms.

Stopping Leaks from Tank to Bowl

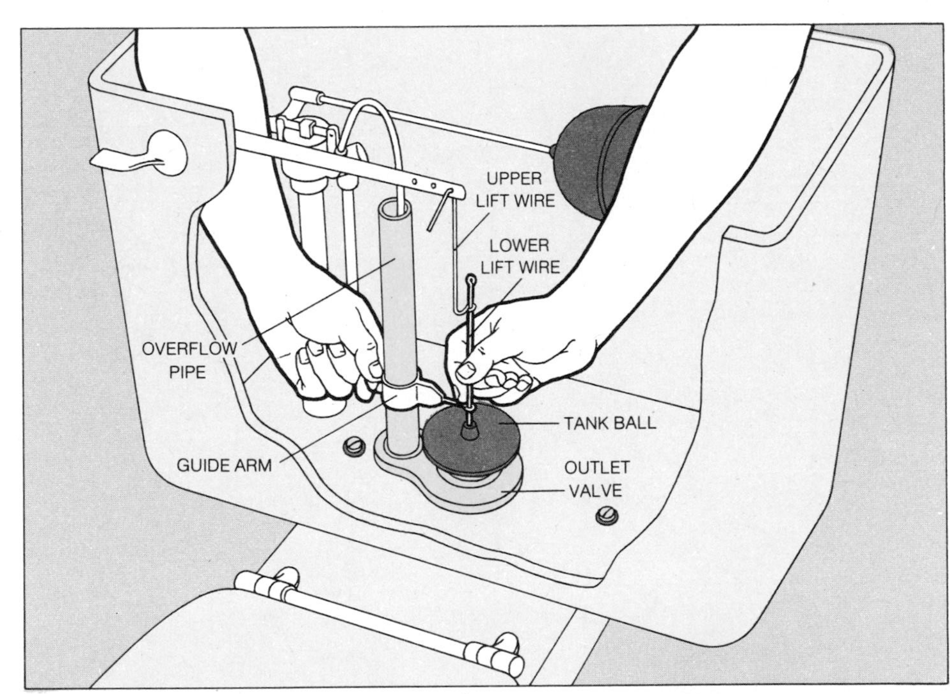

1 Adjusting the tank ball. Water seeping past an imperfectly seated tank ball is the commonest cause of a continuously running toilet. Shut off water to the tank and remove the lid. Flush the toilet and watch the tank ball as it drops with the water level. If it does not fall straight into the outlet valve—the large opening at the bottom of the tank—loosen the thumbscrew that fastens the guide arm to the overflow pipe. Reposition the arm and the lower lift wire so the tank ball will be centered directly over the outlet valve. Straighten both of the lift wires if necessary.

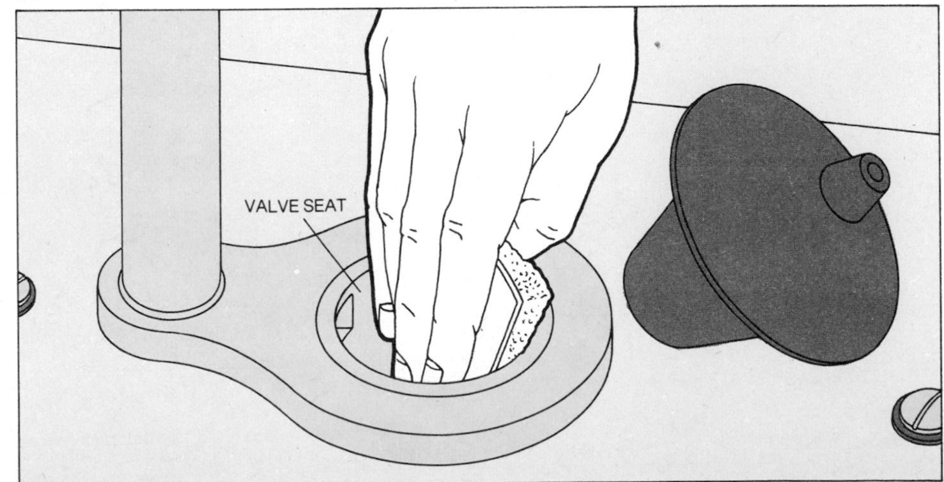

2 Cleaning the tank ball and valve seat. If the tank ball seems to seat properly but water still escapes, the problem may be a build-up of mineral deposits. Unscrew the ball counterclockwise from its lift arm and wash it with warm water and detergent. If the ball is damaged or feels mushy, replace it; an improved type is shown opposite, top. Before replacing the ball, gently scour the seat, or rim, of the outlet valve with fine steel wool or a soap-impregnated plastic cleansing pad *(right)*.

Installing a hinged flapper ball. A flapper tank ball, which has no guide arm and lift wires, is easier to install and less prone to misalignment than the conventional type. The model shown is recommended by most plumbers because it has a semirigid plastic frame that keeps the ball from being jostled out of position by outrushing water.

Drain the tank and remove the old guide arm and lift wires. Slip the collar of the frame to the bottom of the overflow pipe, align the ball over the outlet valve and tighten the thumbscrew on the collar. Hook the chain from the ball through a hole in the trip lever directly above, leaving about ½ inch of slack. Turn the water on, flush the toilet and see if the tank drains completely. If it does not, lessen the slack or move the chain one or two holes toward the rear of the lift arm.

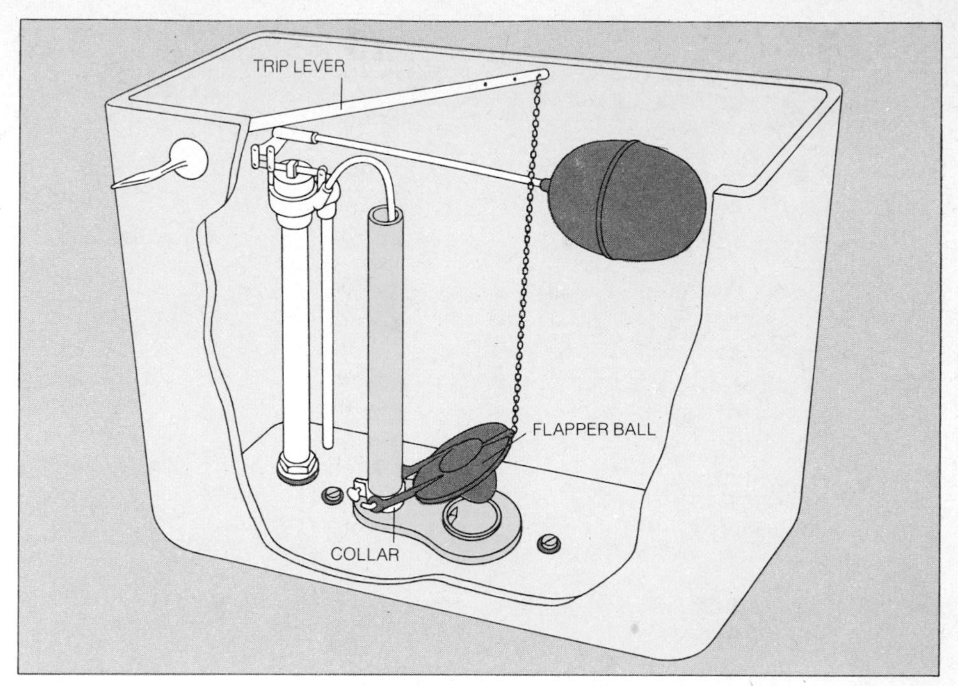

Tightening the handle and trip lever. A loose linkage between the tank handle and the trip lever will cause an incomplete or erratic flush cycle. Use a screwdriver or wrench to tighten the handle on its shaft. If this does not solve the problem, remove the tank lid and tighten the lift lever on the inside of the handle shaft. The model at right has a bracket that lifts the lever; use an adjustable wrench to secure the retaining nut so the bracket does not wobble but still moves freely when the handle is turned. With the wrench or pliers, tighten the trip-lever setscrew against the flattened surface of the handle shaft.

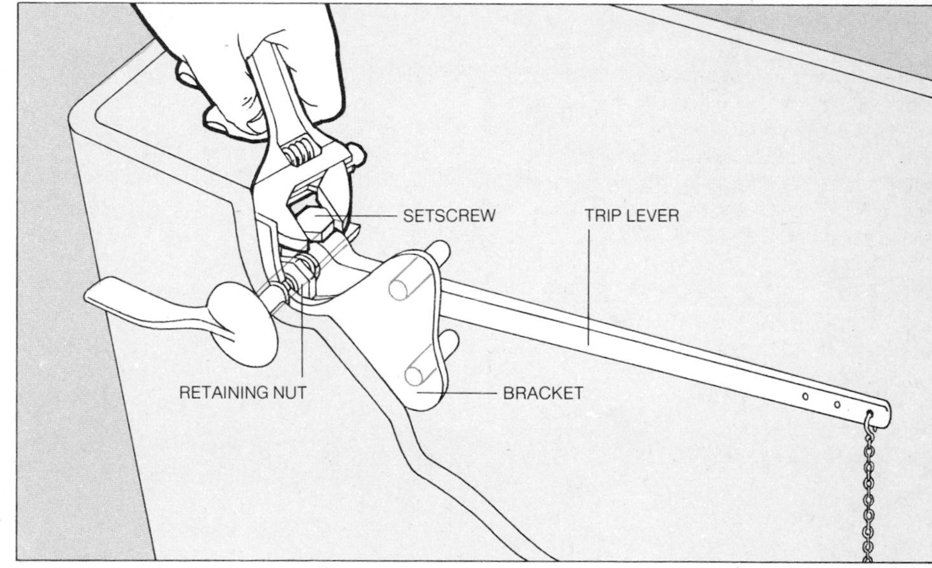

Adjusting the water level. When water cascades into the overflow pipe, the likeliest cause is a defective float ball or one that does not rise high enough to actuate the valve that cuts off incoming water. Unscrew the float ball counterclockwise and shake it; if water has leaked into the float, replace it. If it is sound, grasp the float rod with both hands *(right)* and bend it ½ inch downward. (If you find this difficult, unscrew the rod with pliers and bend it slightly over a rounded surface.) Flush the toilet. The water should stop rising about ½ inch below the top of the overflow pipe. If the water does not reach this level, bend the rod upward.

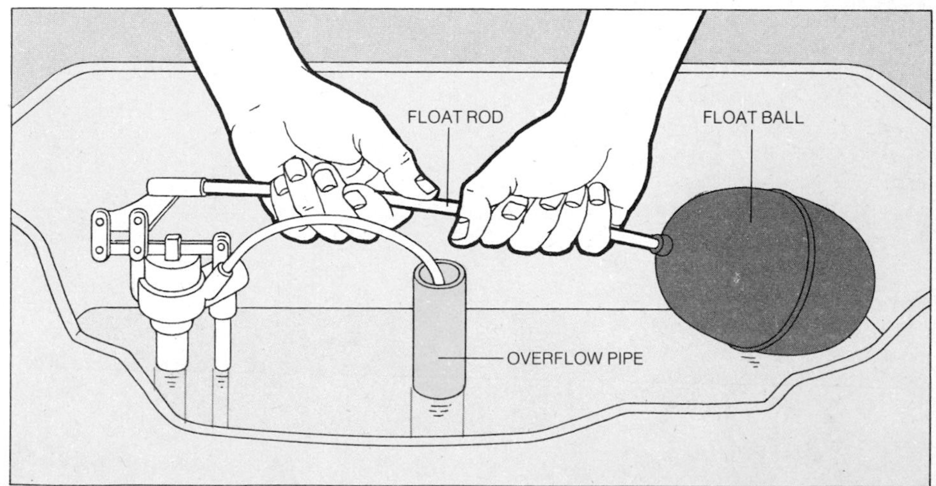

Stopping Ball-Cock Leaks

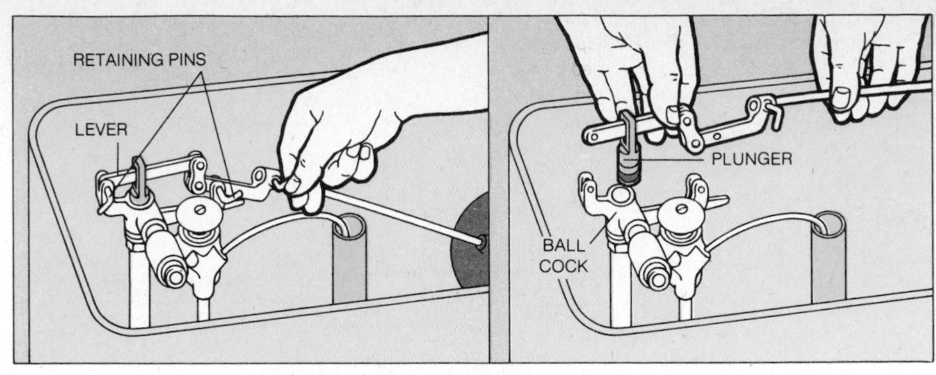

1 Removing the plunger. Check the action of the ball cock by lifting the float rod as high as it will go. If incoming water does not stop, the plunger washers probably are worn. Turn off the water. Unfasten the two retaining pins that hold the float assembly and, with your fingers or pliers, pull the plunger up and out of the ball cock.

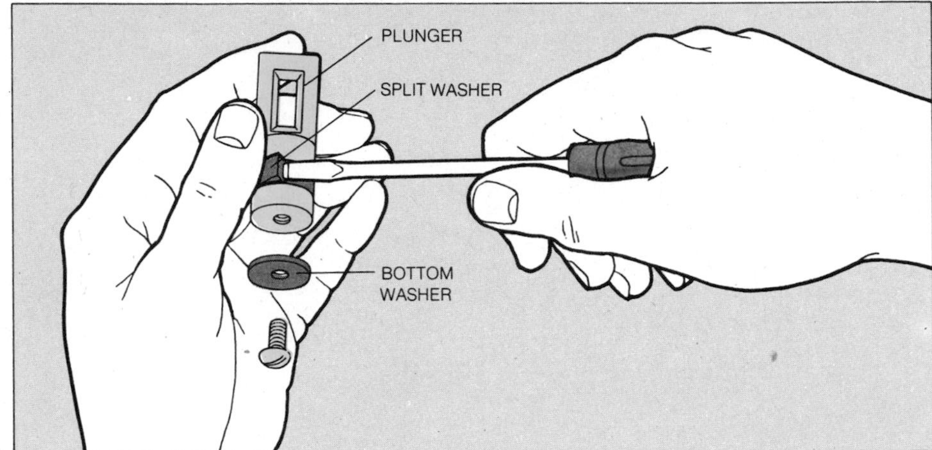

2 Replacing the washers. The plunger has two washers. Pry out the split leather washer (sometimes called a packing washer) from its groove on the side of the plunger. Use a screwdriver to remove the other washer from the bottom of the plunger. Scrape any residue or mineral build-up from both grooves with a penknife or the tip of a small screwdriver, taking care not to scratch or gouge the metal. Install new washers. Larger prepackaged kits of assorted washers often include some for a ball cock. If you have none, ask for a ball-cock washer kit—or, for a perfect fit, take the plunger to your plumbing-supply dealer.

Replacing a Ball Cock

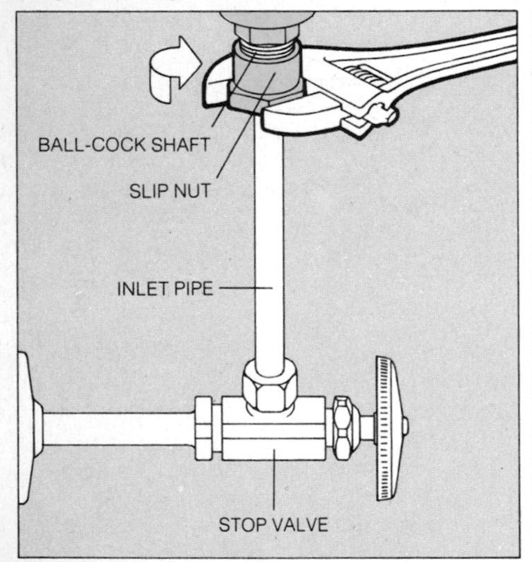

1 Starting the disassembly. A ball cock that still leaks after the plunger washers are renewed should be replaced—further repairs are impractical. Washers and gaskets that form a watertight seal on the tank should also be replaced; these are supplied with the new ball cock. Turn off the water and flush the tank. Sponge out the remaining water. With an adjustable wrench, unscrew the slip nut on the underside of the tank that holds the inlet pipe to the ball-cock shaft.

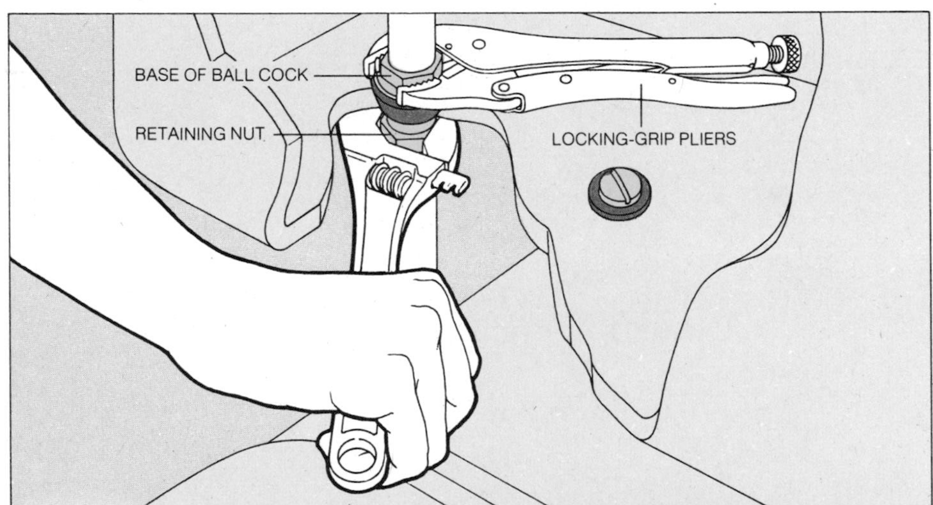

2 Removing the ball cock. Inside the tank, remove the float mechanism from the top of the ball cock and attach locking-grip pliers to the bottom of the ball-cock shaft. The pliers will wedge against the side of the tank and free your hands. Returning to the underside of the tank, use an adjustable wrench to unscrew—counterclockwise —the retaining nut that holds the ball-cock shaft to the tank. Use firm but gentle pressure to avoid cracking the vitreous china of the tank. If the nut resists, soak it with penetrating oil for 10 or 15 minutes and try again. Once the nut is removed, lift the ball cock out of the tank.

3 Installing the new ball cock. The ball cock shown here fills faster, is quieter and has a diaphragm valve instead of a plunger, but is installed in the same way as conventional models. Insert a new washer in the slip nut and place the slip nut washer over the pipe. Put the shank of the ball cock through the gasket and through the hole in the tank. Screw the retaining nut onto the ball-cock shaft. Grip the base of the ball cock with locking pliers and, under the tank, tighten the retaining nut against the tank. Screw the slip nut to the bottom of the ball-cock shaft. Attach the float rod and ball, and the overflow pipe.

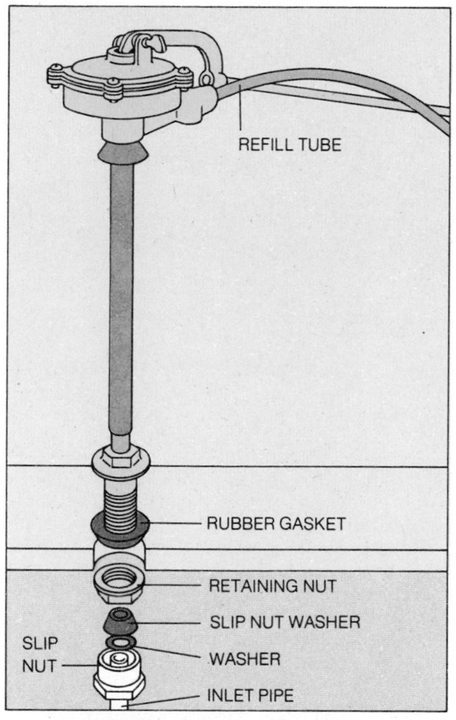

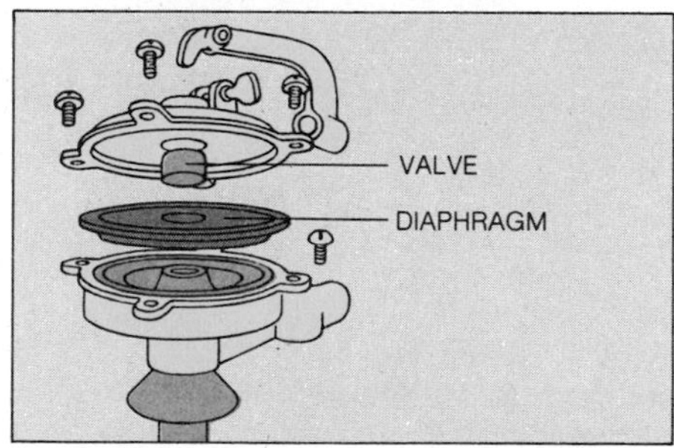

4 Servicing a diaphragm ball cock. If your water supply contains a high concentration of minerals or other impurities you should occasionally clean the moving parts of a diaphragm ball cock. Remove the four screws in the top of the ball cock, lift out the parts and clean them—and the inside cavity—with a plastic scouring pad. After several years of use you may need to replace the diaphragm valve. These parts are available as an inexpensive kit from your plumbing-supply store.

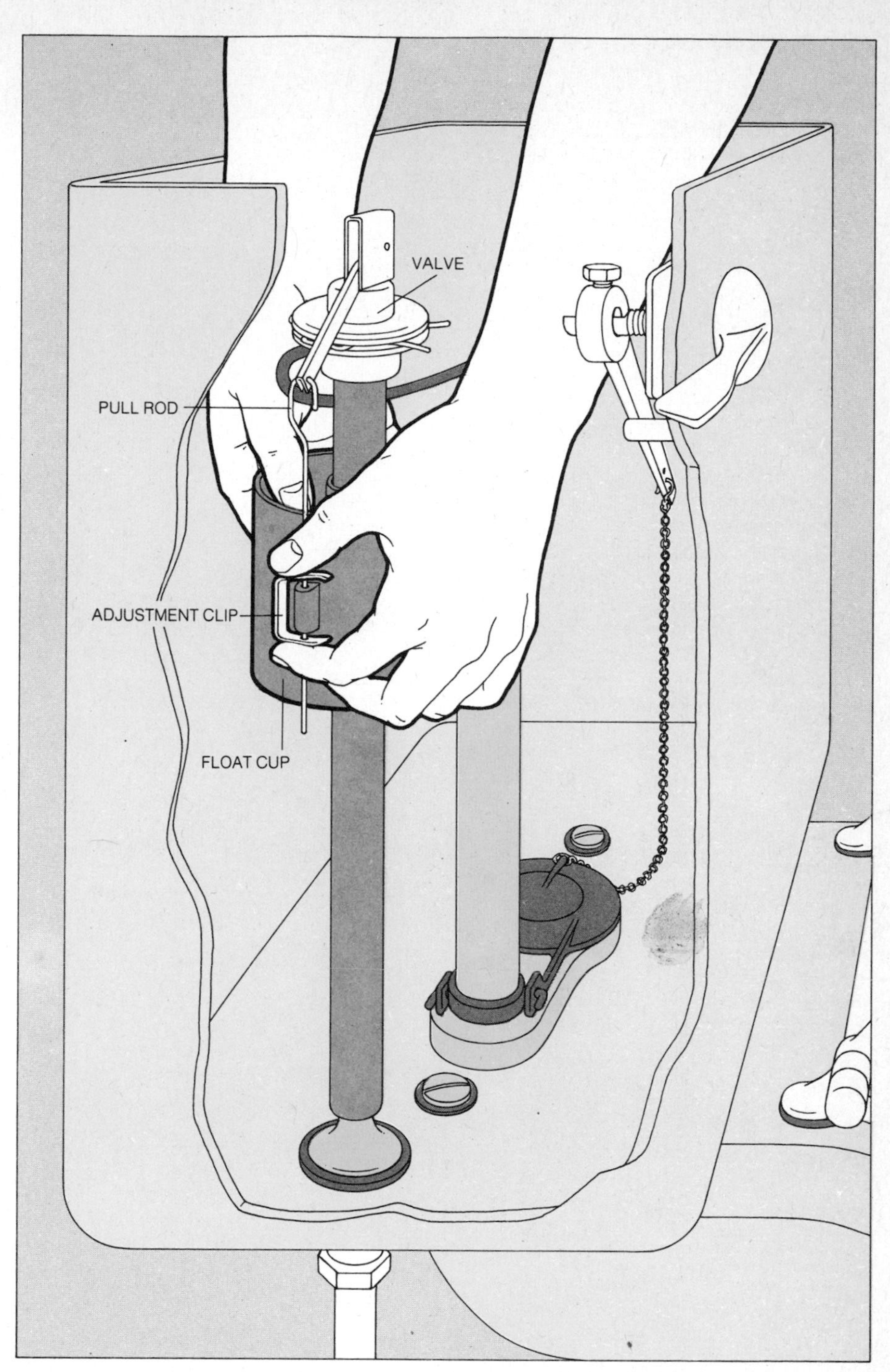

A New Type of Ball Cock

Installing and adjusting a float cup. Many homeowners prefer to replace a worn ball cock. To install a float-cup ball cock, follow the procedure for a conventional ball cock; the fittings are identical. The devices come in several heights, however, so measure the tank depth before buying one. To raise the water level in the tank, simply pinch *(drawing, above)* the spring clip on the pull rod and move the cup higher on the shaft. To lower the level, move the cup down.

Stopping Leaks at Bolts and Gaskets

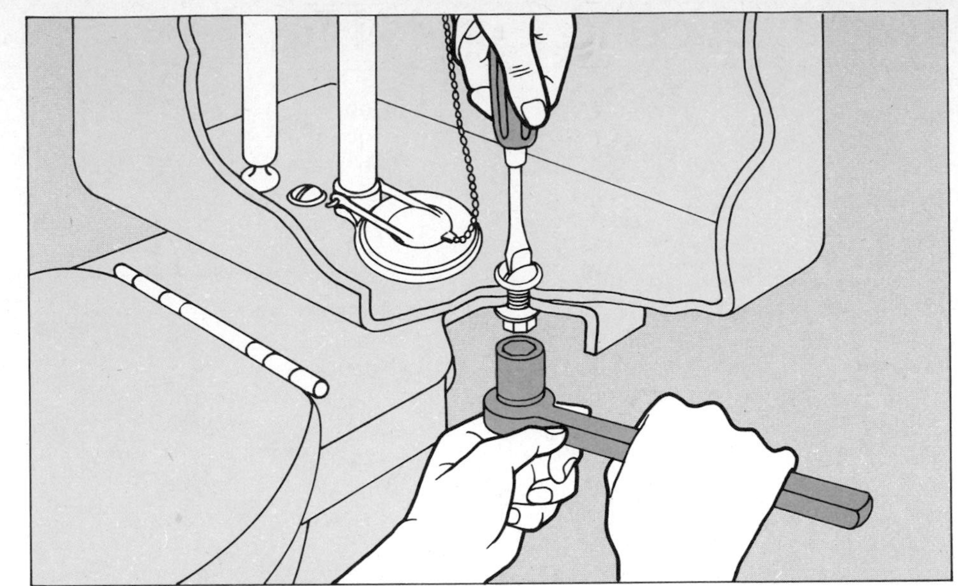

Tightening the hold-down bolts. Seepage around the two bolts that hold the tank to the toilet bowl is often mistaken for condensation. To make a sure diagnosis, pour laundry bluing into the tank and hold a piece of white tissue over the tips of the bolts. If the tissue turns blue, you have a leak. In most cases tightening the bolts will stop the leaks. Drain the tank and hold the head of the bolt—or have a helper do it—with a screwdriver. Use an adjustable wrench, or better, a socket wrench with an extra-deep socket, to tighten the nut below the tank. Caution: Do not overtighten or the brittle finish of the tank will crack. If the leak persists, you will have to remove the bolts and replace their washers.

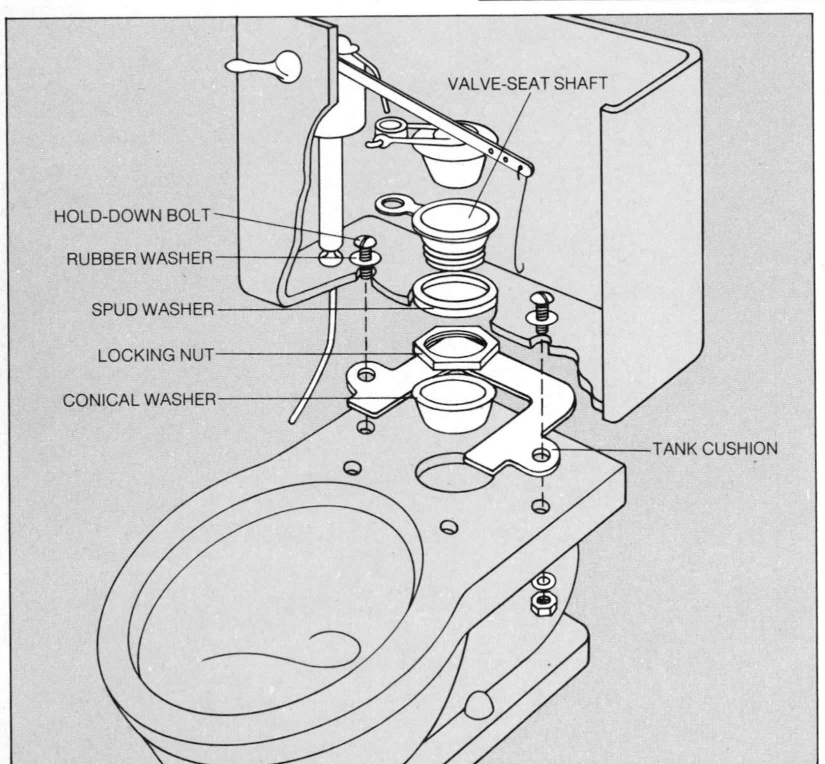

Replacing flush-valve washers. If a leak occurs around the flush valve—the large opening between the tank and the bowl—you must remove the tank. After draining the tank, unscrew the two hold-down bolts and disconnect the supply pipe to the ball cock as explained on the preceding pages. Carefully lift the tank upward and off the bowl, and set it on its back. Remove the large locking nut on the threaded valve-seat shaft that protrudes from the bottom of the tank. Pull the shaft into the tank and replace the spud washer on the shaft. Also replace the large conical washer that covers the threaded shaft of the flush valve. Remount the tank.

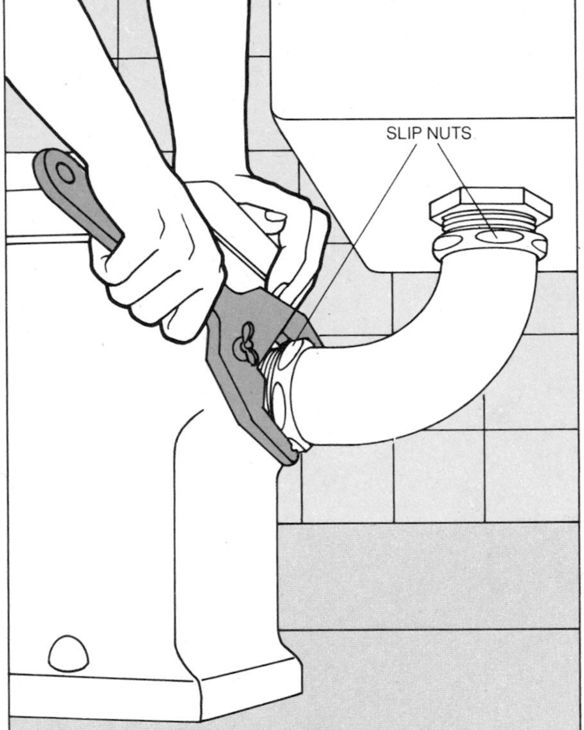

Leaks from wall-mounted tanks. Older toilets have a tank suspended on the wall and connected to the bowl by an exposed L-shaped pipe. Leaks often occur around the pipe fittings because of shifts in wall or floor. Use a larger pipe wrench or a spud wrench to unscrew, counterclockwise, the large slip nuts at the tank and the bowl. Wrap the exposed threads with self-forming packing like that used on faucets *(page 35)* and retighten the slip nuts.

Sweatproofing a Toilet Tank

Condensation of moisture on the outside of toilet tanks is more than a mere annoyance. It encourages mildew. Water that seeps onto the wall or drips from the tank to the floor may cause tiles to loosen or wood to rot.

Tanks sweat because cold water inside the tank cools the porcelain surface, so that moisture in the warm air of the room condenses on it. The housewife's usual panacea, a cloth jacket that fits over the tank, quickly becomes saturated itself and does little to stop the dripping.

Unless the tank water is downright cold—below 50°—condensation usually can be stopped by lining the tank with a waterproof insulating material such as foam rubber. Liners are available at some plumbing-supply stores, or you can buy sheets of foam rubber ½ inch thick at a department store or upholstery shop. Cut it to size with scissors.

If your incoming water supply is often colder than 50°, the only sure way to stop condensation is to raise the temperature of the tank water. This is done by installing a temperature valve, which mixes hot water with the cold water supplying the tank. A hot water line is generally nearby at a lavatory or tub.

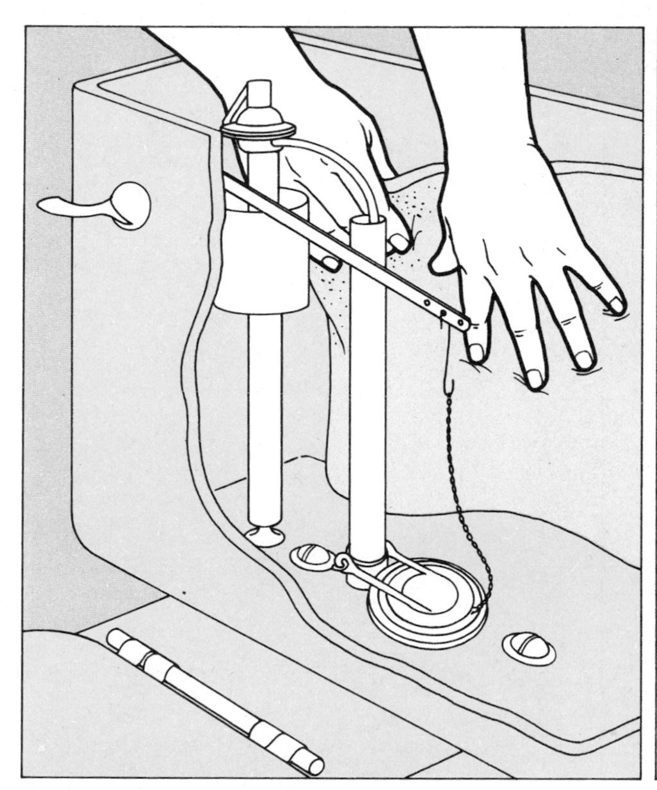

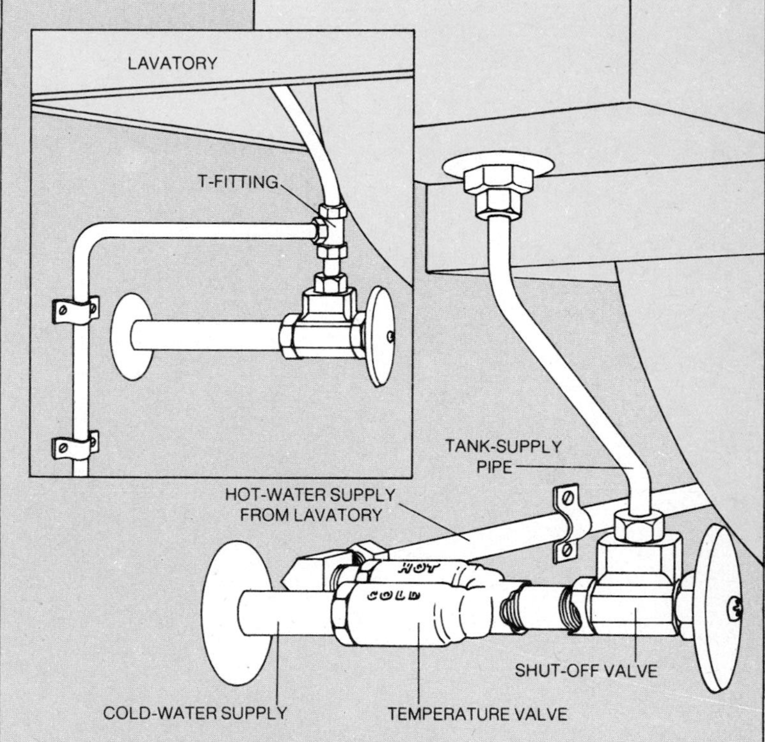

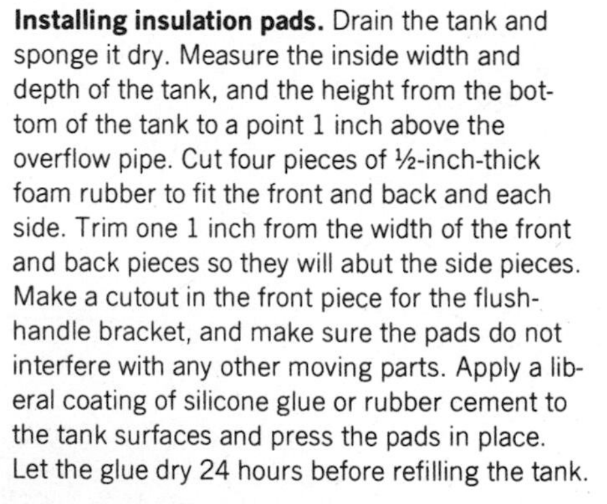

Installing insulation pads. Drain the tank and sponge it dry. Measure the inside width and depth of the tank, and the height from the bottom of the tank to a point 1 inch above the overflow pipe. Cut four pieces of ½-inch-thick foam rubber to fit the front and back and each side. Trim one 1 inch from the width of the front and back pieces so they will abut the side pieces. Make a cutout in the front piece for the flush-handle bracket, and make sure the pads do not interfere with any other moving parts. Apply a liberal coating of silicone glue or rubber cement to the tank surfaces and press the pads in place. Let the glue dry 24 hours before refilling the tank.

Installing a temperature valve. Shut off the water at the main cutoff. Disconnect the tank supply pipe and unscrew the toilet shutoff from the supply pipe emerging from the wall. Screw the "cold" inlet of the temperature valve onto the supply pipe. Install a threaded nipple between the valve outlet and the shutoff inlet. Reconnect the inlet pipe to the tank with a flexible connector *(page 99)*. Install a compression T fitting with a ¼-inch takeoff adapter in the nearest hot water line *(inset)*. Run ¼-inch flexible copper tubing from the T and connect it to the "hot" temperature-valve inlet; using a compression adapter. Attach the tubing to the wall with brackets.

Replacing a Toilet Seat

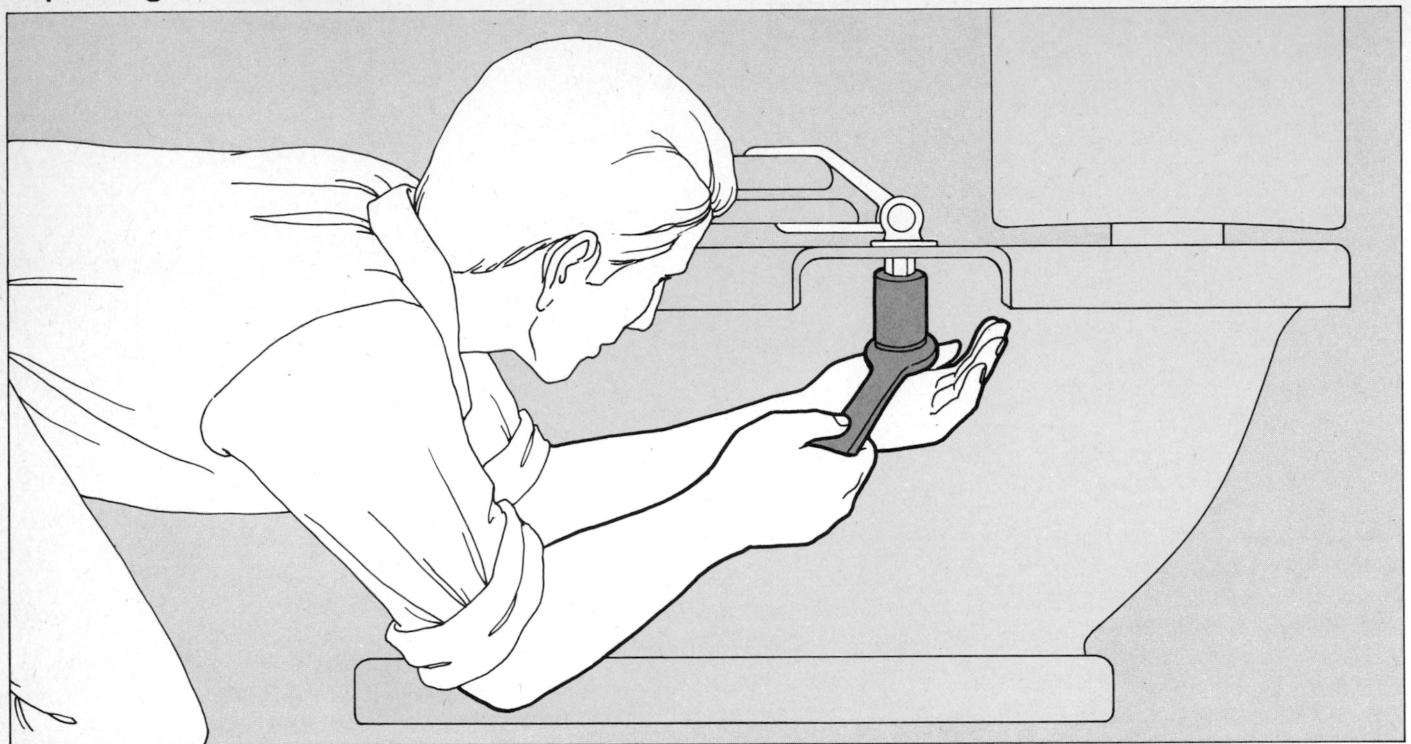

Removing the seat bolts. The two bolts that hold the seat to the bowl generally have smoothly rounded heads that provide no gripping surface and must be removed by unscrewing, counterclockwise, the nuts underneath the bowl. But the nuts usually are recessed in the lip of the bowl and are also hard to grip. If long-nose pliers will not loosen them, try a socket wrench with an extra-deep socket. If metal bolts are so frozen by corrosion that these methods cannot loosen them, try the procedures below.

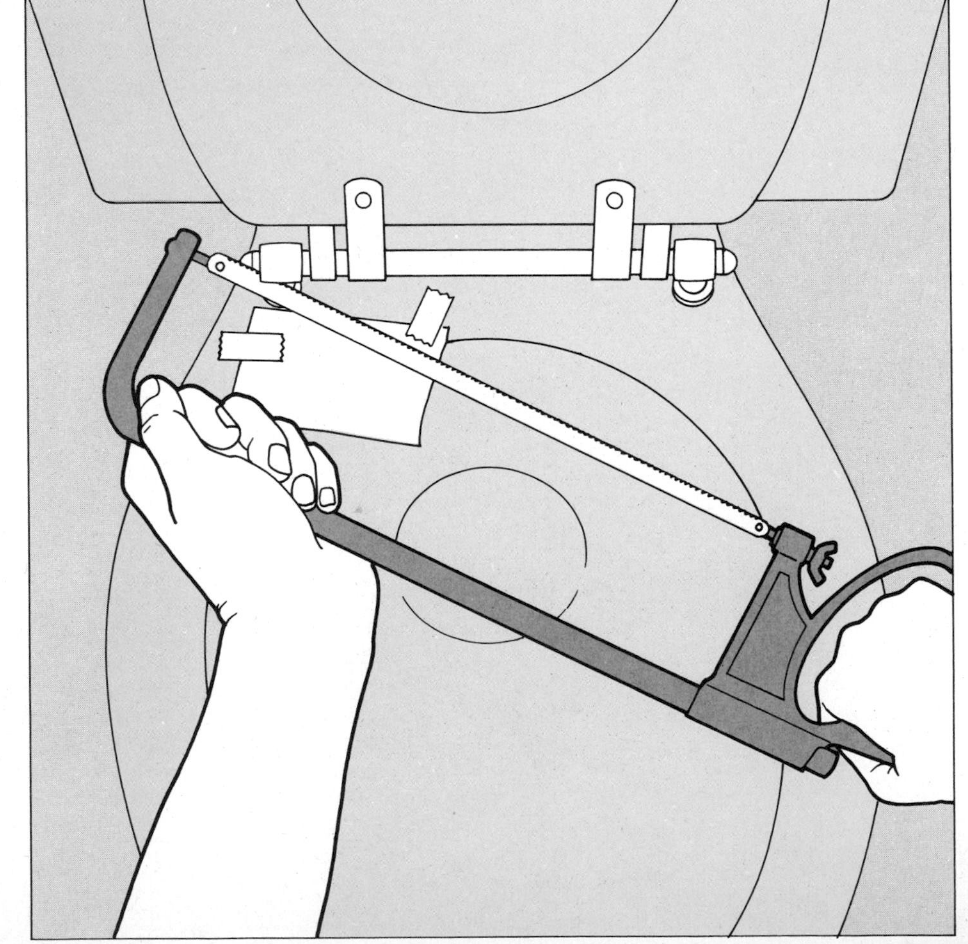

Freeing corroded bolts. Soak the bolts with penetrating oil for 30 minutes or even overnight, then try loosening the nuts once more. Do not use force; the brittle china bowl cracks easily.
If penetrating oil does not free the nuts, use a hacksaw to cut off the boltheads, sawing through their attached washers. Place thin cardboard between blade and bowl to protect the china.

Repairing Pressure-Valve Toilets

To provide the forceful volume of water needed for flushing action directly from a water supply system, rather than from a storage tank, a pressure-flush valve toilet must be connected to a larger-than-usual pipe, 1 or 1½ inches in diameter. It lets water flow at 30 gallons per minute, and this high flow rate can lead to waste or, if parts fail, flooding.

A pressure valve toilet contains three mechanisms: a built-in stop valve to shut off water quickly in an emergency, a handle assembly and the pressure valve itself. Handles and stop valves are essentially the same on all toilets. There are, however, two distinct types of pressure valves, one employing a diaphragm mechanism and the other a piston. The two types can be easily identified by their outside profiles as shown in the drawings at top.

Fix leaks or other malfunctions by replacing worn parts. Since parts made by different manufacturers are not interchangeable, be sure to note the maker's name; it is clearly stamped on the device, usually on top of the pressure valve.

The flow rate of a diaphragm-type pressure valve is regulated by adjusting the stop valve *(page 62)*. The diaphragm automatically regulates the duration of the flush cycle. On a piston-type pressure valve, both the flow rate and the flush cycle are regulated by turning a screw on top of the valve *(page 62)*.

Two types of pressure valves. A diaphragm valve *(below)* is recognizable by its one-piece, rounded cover, which is nearly twice as large as the pipe leading to it. The cover of a piston valve *(right)* is slightly larger than the pipe containing the valve.

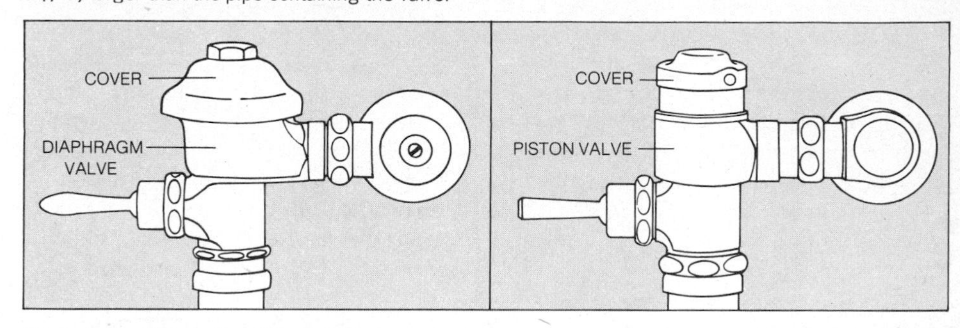

Solving Handle Problems

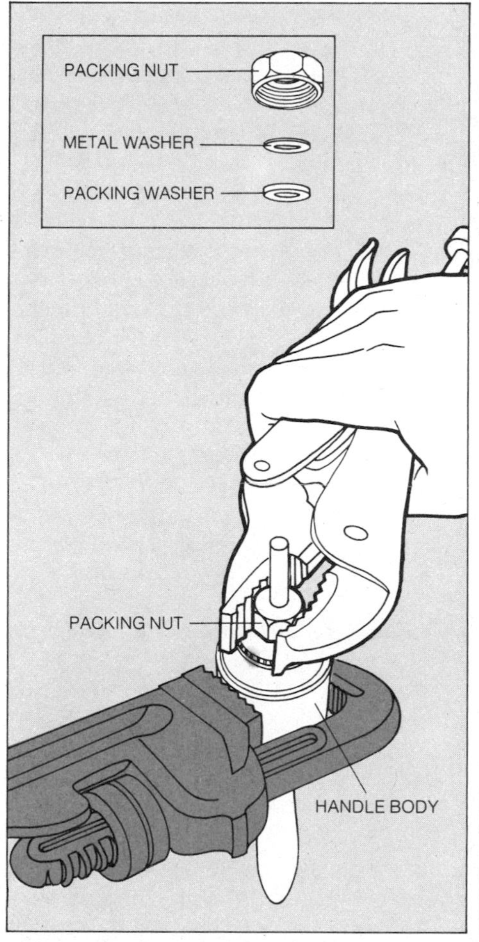

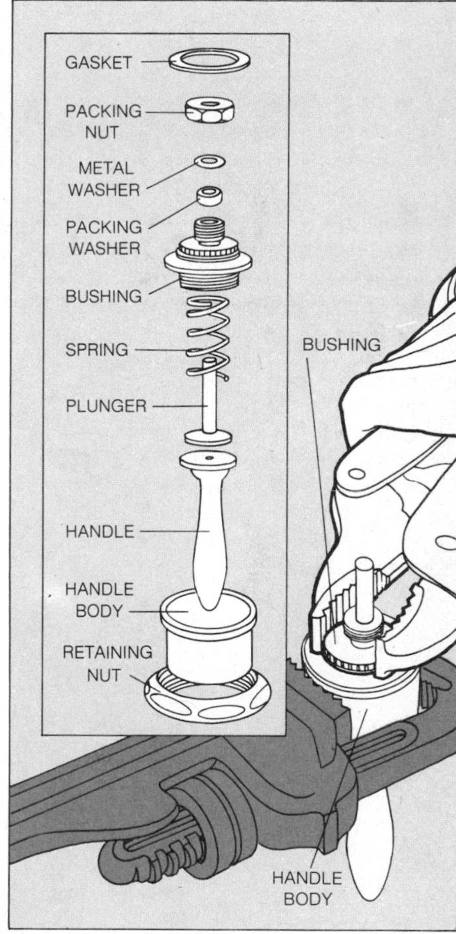

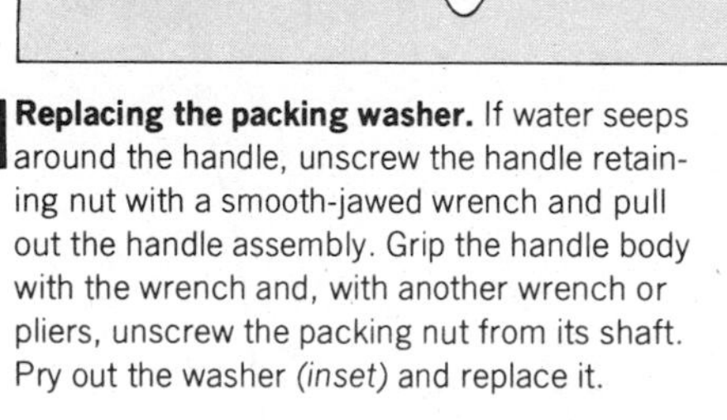

1 Replacing the packing washer. If water seeps around the handle, unscrew the handle retaining nut with a smooth-jawed wrench and pull out the handle assembly. Grip the handle body with the wrench and, with another wrench or pliers, unscrew the packing nut from its shaft. Pry out the washer *(inset)* and replace it.

2 Replacing inner-assembly parts. Persistent leaks or a loose and wobbly handle usually are caused by a worn bushing, spring or plunger inside the handle body. Remove the handle as you would in replacing a packing washer. Wrap the handle body with adhesive tape to avoid marring it. Hold the body in a pipe wrench and remove the inner assembly by unscrewing its bushing with locking-grip pliers. Replace worn parts with new ones, available in a complete kit *(inset)*.

Regulating the Flush Cycle

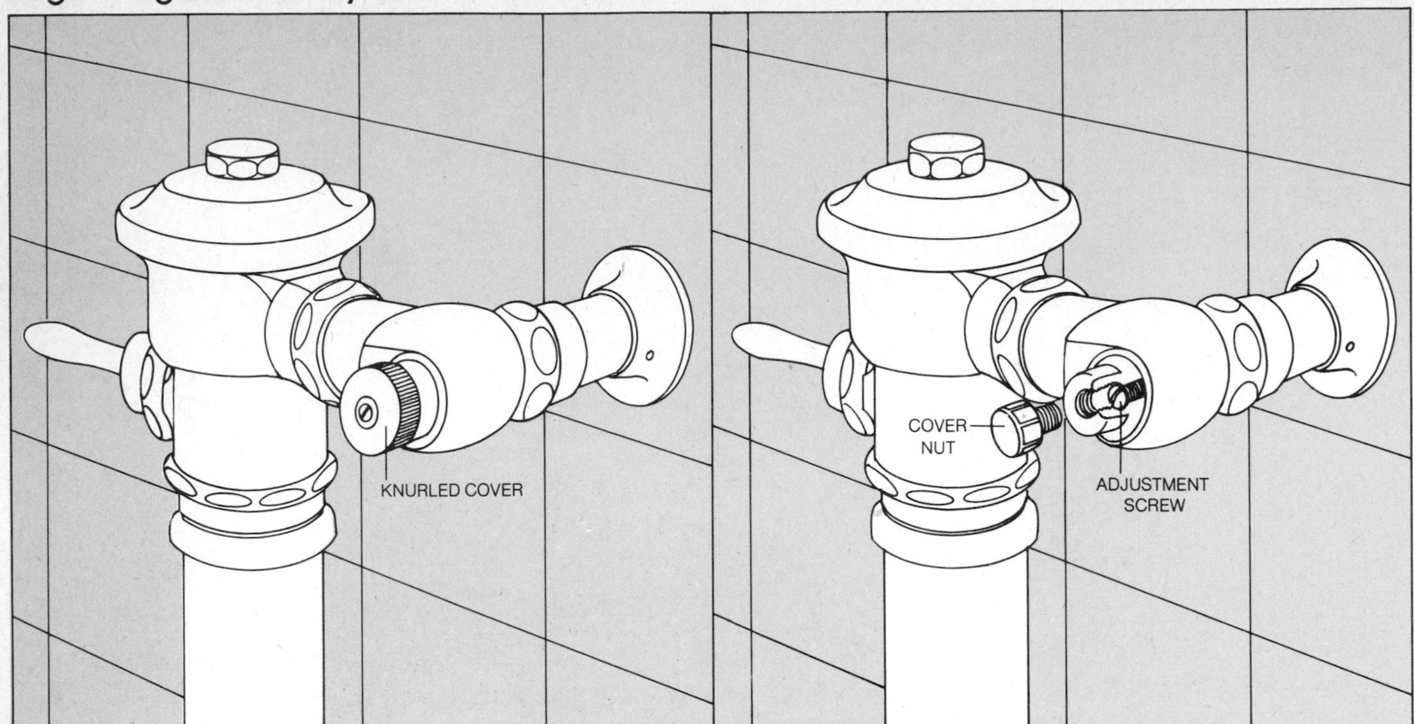

Adjusting stop valves. The shutoff for any pressure valve toilet is located in the large pipe that emerges from the wall. Some have handles like any other stop valve, but many have handles disguised as knurled covers *(above, left)* and others have no handle at all—the valve is operated by turning a screw, and the screw may be hidden by a cover nut *(above, right)*.

To shut off the water, turn the handle or screw fully clockwise. Leave the stop valve fully open on a piston-type toilet. On a diaphragm-type toilet, open the valve only slightly; it regulates flow as described on page 61.

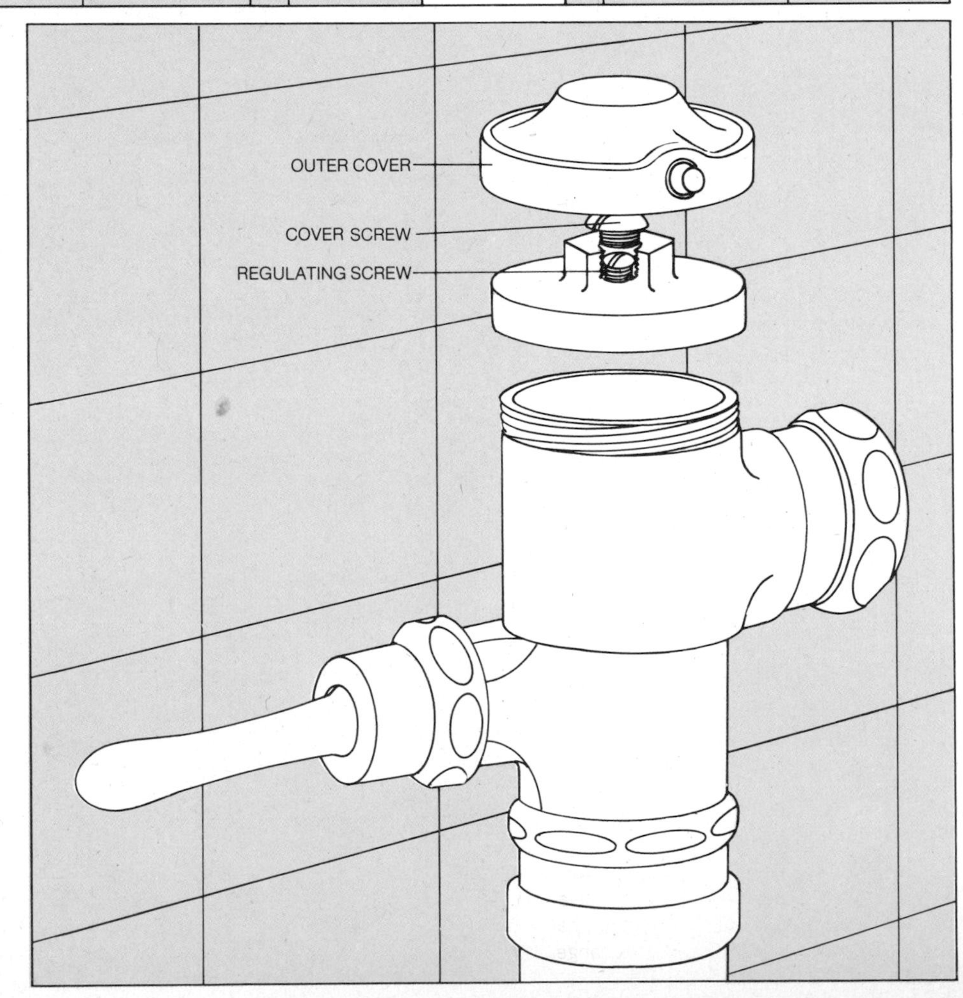

Regulating flow. On a diaphragm-type toilet, open the stop valve one or two turns, setting it so that the toilet flushes quickly without flooding but refills the bowl to the normal level. On a piston-type toilet *(right)*, remove the cap of the piston valve assembly—usually it simply lifts off—and, with a screwdriver, take out the cover screw underneath the cap. Turn the recessed regulating screw fully clockwise, then back it off one or two turns, turning it farther back for a longer flush cycle, forward for a shorter one.

Repairing a Piston Pressure Valve

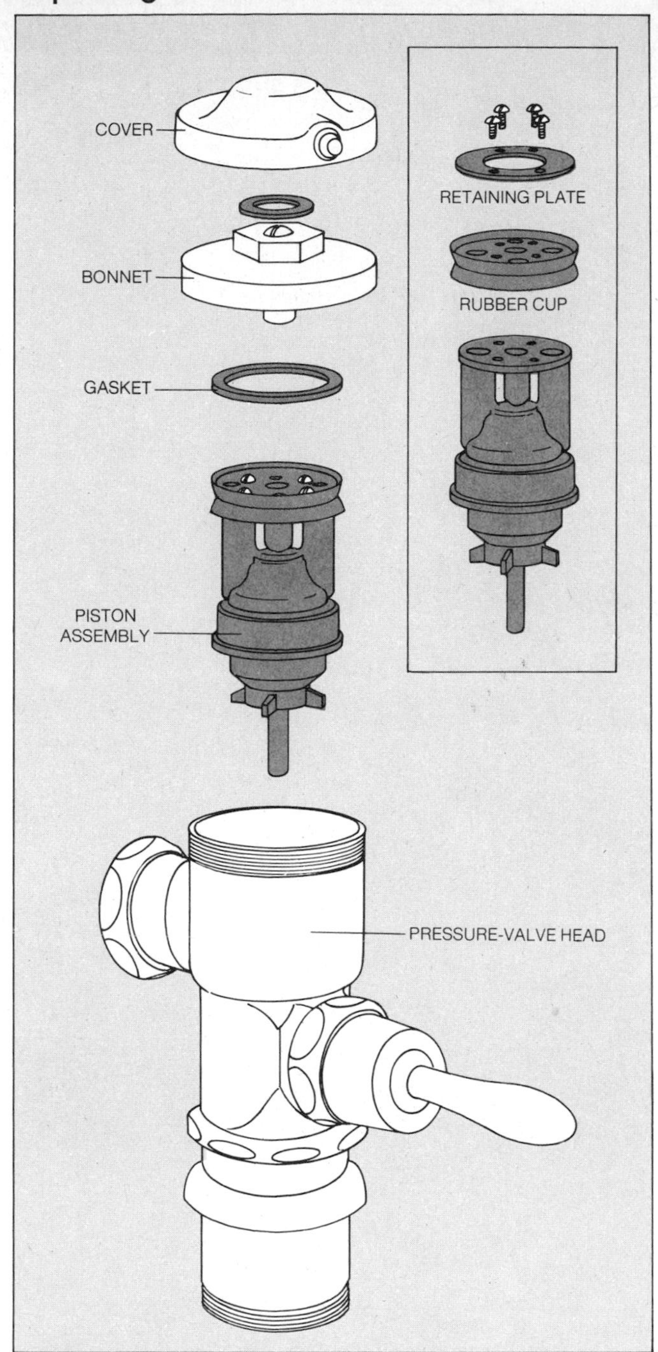

Removing the piston assembly. Shut off the water and lift the cover off the top of the pressure-valve head. With a smooth-jawed wrench loosen the bonnet and then unscrew it the rest of the way by hand. Carefully remove the rubber gasket and pull out the piston assembly.

Only the double-lipped rubber cup at the top of the piston assembly is replaceable. Remove the retaining screws, lift out the retaining plate and install a new cup. Older piston valves may have a leather cup with a single lip. It is interchangeable with the more efficient double-lipped cup.

Repairing a Diaphragm Pressure Valve

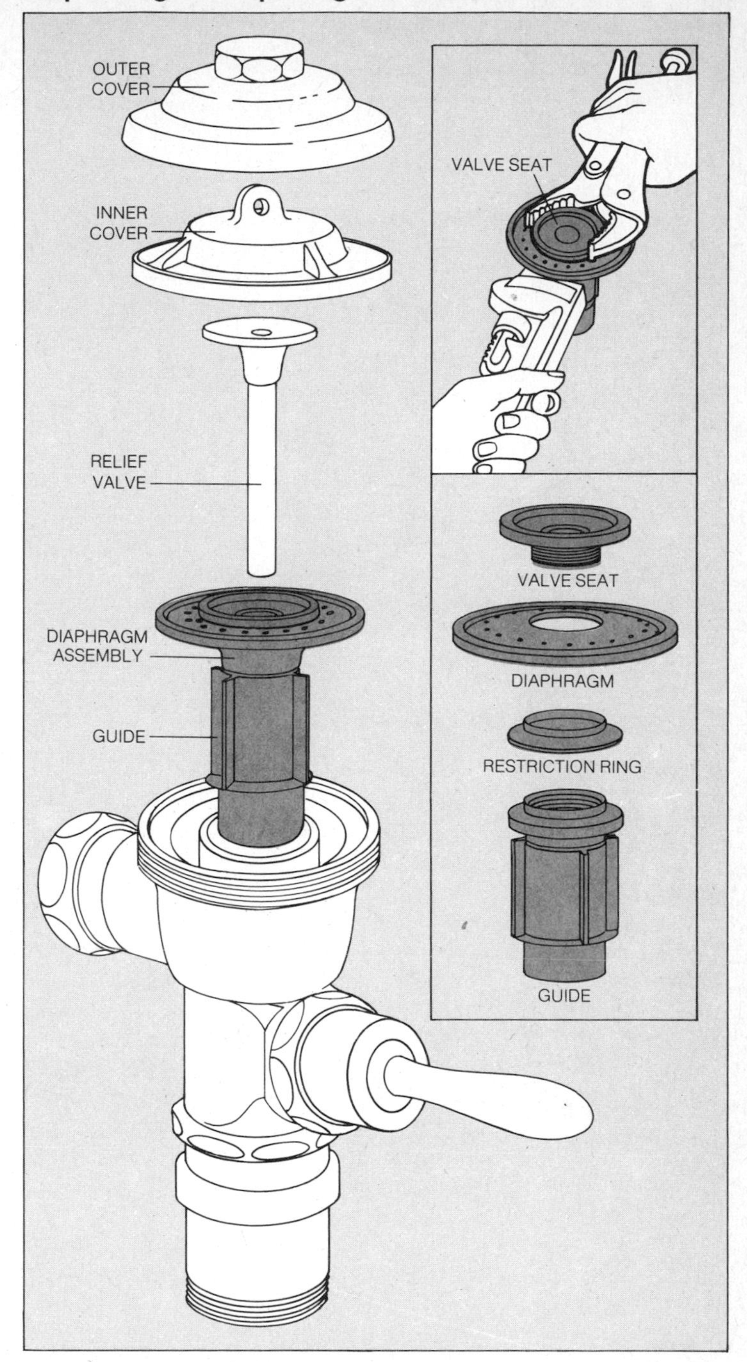

Removing the diaphragm assembly. Shut off the water. With a smooth-jawed wrench, unscrew the outer cover, which is molded in one piece. Lift off the inner cover; if it is firmly seated, insert a screwdriver in the loop on top of the cover and twist gently to break the vacuum. Pull out the diaphragm assembly and remove the relief valve.

To replace internal parts, hold the metal guide of the assembly with a wrench *(inset, top)* and, using another wrench or pliers, unscrew the rubber valve seat from the top center of the diaphragm. Both usually need replacing.

A New Look for Flawed Ceramic

The pretty parts of the plumbing system —ceramic tile, soap dishes and towel bars, and the porcelainized surfaces of fixtures—need repairs as much as the pipes. Gleaming chrome dulls with layers of mineral deposits; ceramic or porcelain enamel is discolored by rust stains —usually because of a dripping faucet that should have been repaired—or by chemicals in water. A hair dryer slips from the hand—and you have a chipped bathtub. Ceramic tiles may even loosen and fall from the wall because of poor installation or the cumulative effects of time and humidity. And one of the most common bathroom repairs—sealing cracks in the joint between the bathtub and the wall—is a repetitive chore because of changes in the weight of the tub as it is filled and emptied again and again.

The chart and instructions on the following pages provide solutions for these problems. In some situations you may have to combine two techniques to get a job done. On page 67, for example, you will find instructions for replacing a broken towel bar. If it turns out that the mounting plate for the new bar does not cover the scar left by the old one, modify the instructions: replace the entire tile by the technique shown opposite, then mount the towel bar on the new tile by the technique on page 67, Steps 2 and 3.

Whatever job you tackle, take practical precautions. Wear goggles when you shatter a tile; wear gloves when you work with powerful cleansers or wet grout. And when surrounded by the brittle surfaces of a bathroom, handle heavy tools with special care—a single slip can add a new repair job.

Cleaning Tiles and Fixtures

Problem	Solution
Tile adhesive on the surface of the tile	Wash it off quickly with a damp cloth. If it has set, scrape it off carefully with a razor blade or window scraper—without scratching the tile—and then wash the tile with paint thinner.
Dirty seams between tiles Cloudy, filmy tiles	Wash with a toothbrush and detergent. Remove light film with detergent. Heavily soiled tiles need to be washed with trisodium phosphate, available in many paint stores (if phosphates are banned in your community, use a heavy-duty household detergent). Dilute 1 teaspoon of the trisodium phosphate in water. For discolored tiles, sprinkle the phosphate on a moist cloth and wipe, then rinse and dry with a clean, soft cloth.
Rough, grainy surface on tiles or enameled bathtub from the lime deposits in hard water.	Dissolve lime deposits by washing with vinegar.
Discolored bathtub	To improve the appearance of an enameled tub, use a mixture of cream of tartar and peroxide. Stir in enough peroxide to make a paste and scrub vigorously with a small stiff brush.
Soiled or discolored toilet bowl	Flush to wet the sides of the bowl, then sprinkle toilet cleaner or chlorine bleach on the wet surfaces. Let it stand a few minutes (longer if badly stained), then brush with a long-handled toilet brush and flush. Never try to strengthen the cleanser with chlorine bleach or add ammonia to either: the mixture will cause a chemical action that liberates toxic gases.
Dirty or sticky chromium faucets	Wash with a mild soap or detergent and polish with a dry, clean cloth. Use vinegar to remove mineral deposits. Do not use metal polishes and cleaning powders: they will damage the plating.
Iron rust stains on sinks or bathtubs	If the fixture is lightly stained, rub with a cut lemon. If seriously stained, use a 5 per cent solution of oxalic acid or 10 per cent hydrochloric acid. Apply with a cloth, leave on only a second and wash off thoroughly. Repeat if necessary.
Green copper stains on sinks or bathtubs	Wash with strong solution of soapsuds and ammonia. If the stain persists, try a 5 per cent solution of oxalic acid.

Replacing a Ceramic Tile

1 Removing the tile. Using a hammer and cold chisel and wearing goggles, smash the center of the tile, then pry out the pieces. If a tile falls out in one piece and can be reused, scrape away the old cement on its back. Remove all loose or uneven cement from the tile bed.

2 Resetting the tile. Apply tile adhesive to the back of the tile and holding it by its edges, set it into its place so that it is even with the adjacent tiles. If the tile does not have little spacer lugs on two sides, place toothpicks in the joint to space it. Let the tile adhesive dry overnight.

3 Applying the grout. Use already-mixed tile grout to fill and seal the joint spaces around the new tile. First force the grout into the joints with the tip of your finger, then use a moist cloth or window squeegee to remove the surplus grout from the tiles on either side.

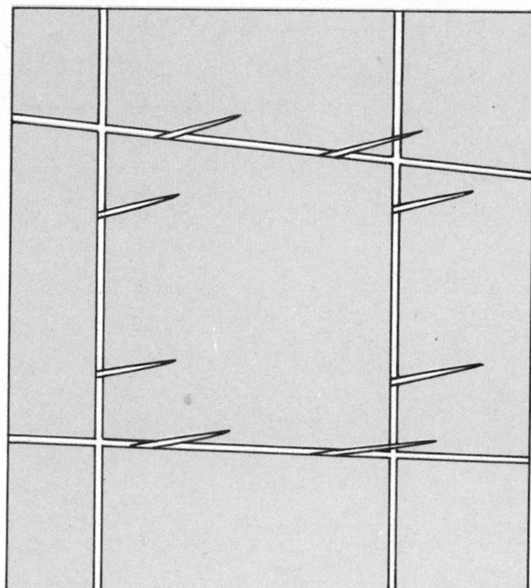

Shaping a Tile to Fit

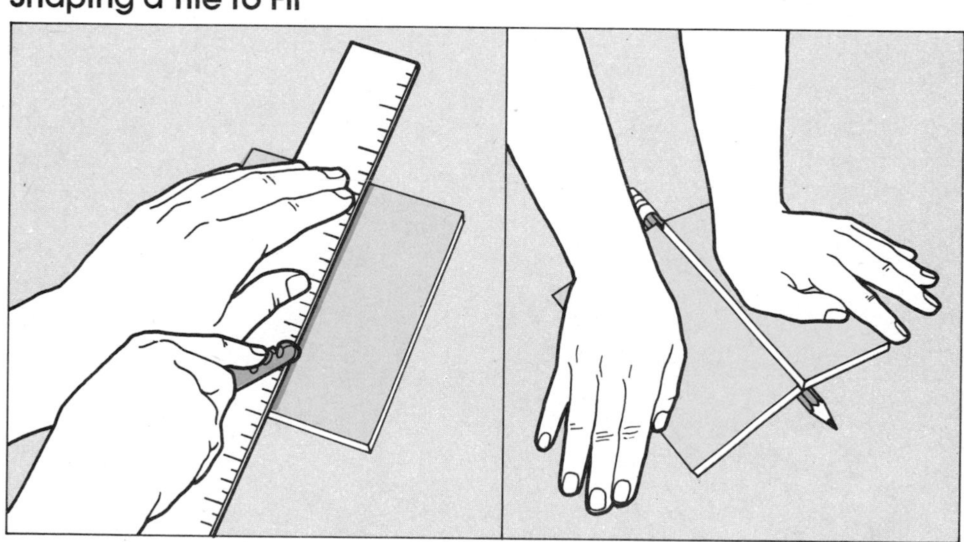

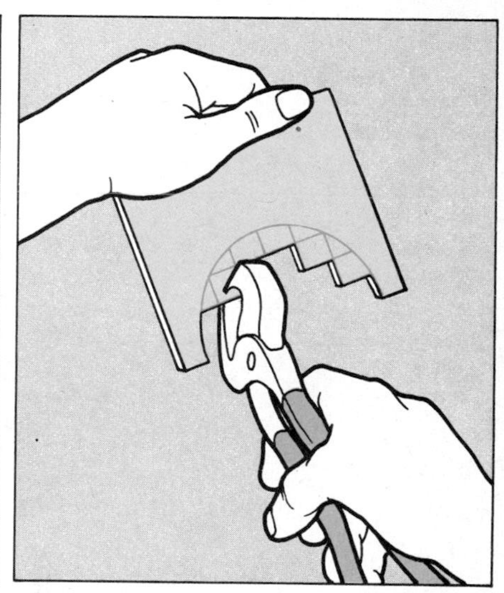

A straight cut. If a tile must be cut to fit into a narrow rectangular space alongside a fixture, outline the part to be removed with a grease pencil. Score the line with a glass or tile cutter: press firmly on the cutter and pull it along the pencil line with a smooth motion *(above)*. To break the scored tile in two, place a pencil on a flat surface and position the tile on the pencil so that the scored line is directly over it. Press down equally on both sides of the line until the tile snaps *(center)*.

A curve cut. To fit tile to a curved fixture, outline the excess part of the tile in grease pencil, and with a glass cutter score a grid over the area to be removed. Using pliers or tile nippers, chip away the grid area. File the edges smooth.

Sealing the Joint between Wall and Tub

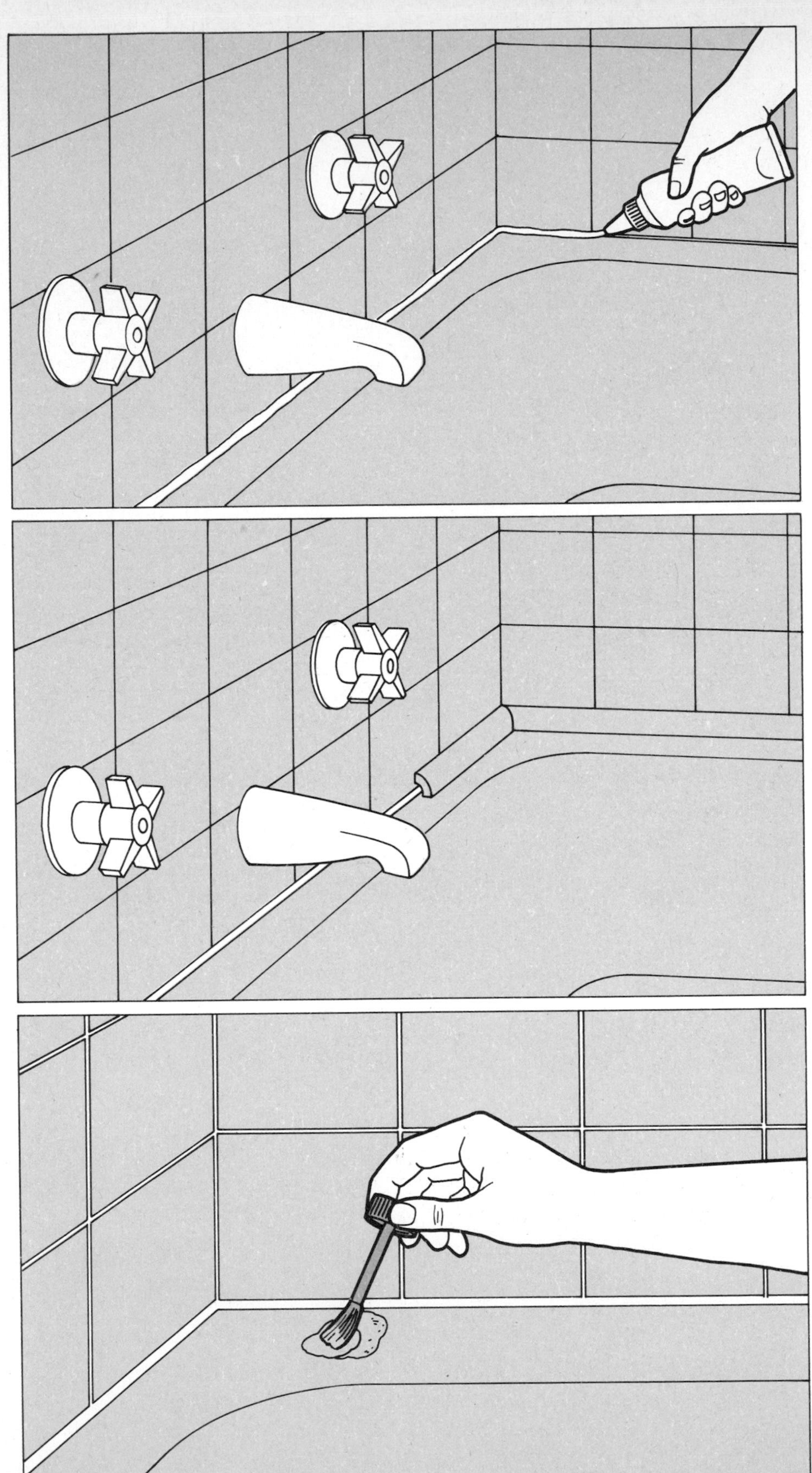

Caulking the joint. The simplest way to fill the joint between wall and bathtub is with flexible waterproof caulking compound. Slowly squeeze the compound out of its tube, using as steady and continuous a motion as possible. Wait at least 24 hours before using the bathtub.

Edging tiles. If caulking will not stay in the wall-bathtub joint, apply quarter-round ceramic edging tiles, which are available in kits. These tiles are easily attached around the entire rim of the tub with a caulking compound.

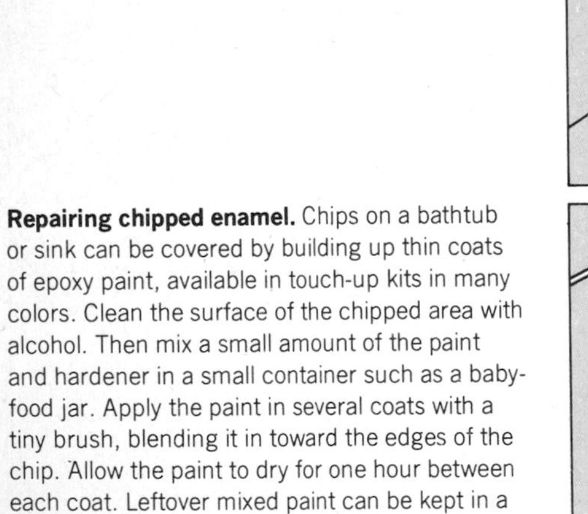

Repairing chipped enamel. Chips on a bathtub or sink can be covered by building up thin coats of epoxy paint, available in touch-up kits in many colors. Clean the surface of the chipped area with alcohol. Then mix a small amount of the paint and hardener in a small container such as a baby-food jar. Apply the paint in several coats with a tiny brush, blending it in toward the edges of the chip. Allow the paint to dry for one hour between each coat. Leftover mixed paint can be kept in a refrigerator for as long as 72 hours.

Ceramic Accessories

Broken soap dishes, towel racks and grab bars are replaceable—but as a rule should not be duplicated. In most cases, it is easier to use one that is attached differently.

Most accessories are originally set into tile walls with portland cement, which is messy to handle. It is therefore best to replace a grab bar or towel rack with a type that can be screwed to the wall. A light soap dish can be simply applied with tile adhesive; select one without a grab handle, so that you will not be tempted to pull yourself up on it.

Replacing a Soap Dish

1 Removing the dish from the wall. With a utility knife, score the grout around the soap dish. Protect the adjacent tiles by covering their edges with masking tape. Then, wearing goggles, lightly hammer the broken parts of the dish to loosen them. Set a cold chisel in the groove made by the knife and tap it to force out the dish. Remove the old grout and tile adhesive.

2 Replacing the dish. Select a replacement dish that will take up exactly the same number of tile spaces as the old one. Attach the replacement by applying a coat of tile adhesive to the back of the soap dish, then hold it in place with masking tape until the adhesive sets. Wait at least 24 hours for the cement to dry thoroughly, then seal the joints with grout.

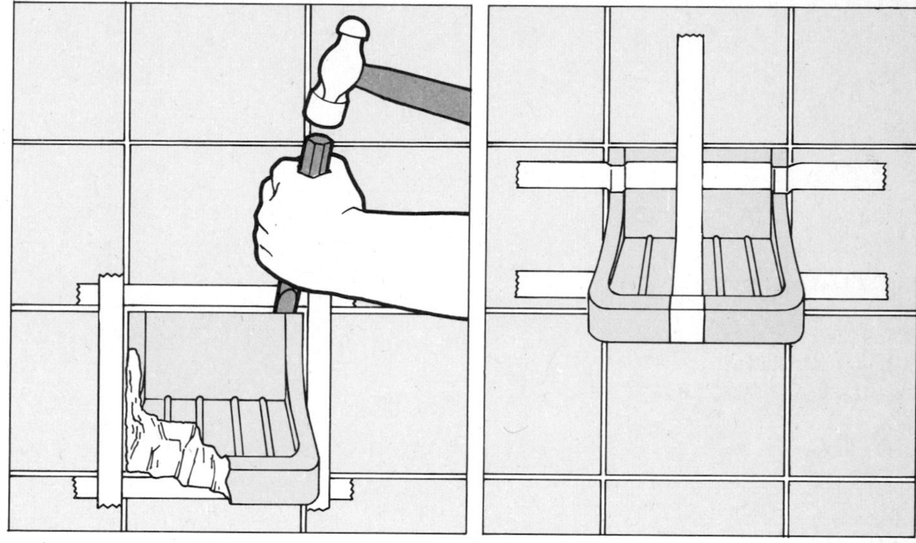

Replacing a Grab Bar or Towel Rack

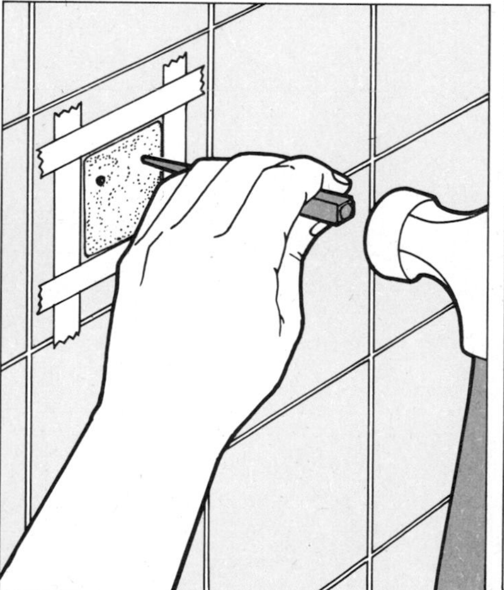

1 Removing the old accessory. If a grab bar or towel rack breaks, do not gouge out the cement or remaining ceramic piece, but try to get it as flush as possible with surrounding tiles. Protect the tile with masking tape and, wearing goggles, use a hammer and cold chisel, then the sanding attachment on an electric drill to remove as much protrusion as possible. At the inner edges of the tape, sand the surface by hand.

2 Drilling through tile. Position the plate for your new rack or bar so that it conceals the old cement bed. Use a metal punch and hammer to knock small chips off the surface of the tile so that the drill bit will not skid, then drill holes for the plate with a masonry bit.

3 Fastening the mounting plate. Most contemporary bathroom accessories have "hidden" plates; that is, plates that are attached to the wall with screws and are concealed by the final assembly. If the tile is set in portland cement, insert plastic or lead anchors for the mounting-plate screws *(above)*; if you are fastening the unit to a dry wall, use toggle bolts or hollow-wall anchors. Complete the assembly of the unit according to the manufacturer's directions.

SOLDER

3 Working with Pipe

Soldering a joint. Copper pipe—the most common type of water supply pipe—is strongest when joined by soldered "sweat" joints. The pipe and fitting, here a 45° elbow, are heated with a propane torch *(bottom)* until hot enough to melt a strip of solder *(top)* on contact. The solder will be held to the edge of the fitting until a solid ring of molten metal appears around the entire joint, as it does on the completed left side of the fitting.

It is not often that you are forced to cut out a section of pipe and insert a new length—only on that rare occasion when a supply line freezes and bursts, a joint springs a leak or a drain becomes permanently clogged. More often you may decide you want to add an entirely new branch of piping as changing family needs demand additional fixtures and appliances. But whether from choice or necessity, the job is easier than you may think, for modern materials and easy-to-master professional techniques make these chores much less daunting than they at first appear.

New materials have revolutionized plumbing. No matter what kind of piping you already have—copper, steel, cast iron, plastic or even old-fashioned lead—you can generally make a choice among several materials to use for a repair or a new branch, since there is an array of adapter fittings *(page 71)* that allows any one type to be joined to almost any other. Make the job as simple as possible by selecting the easiest pipe material to use for the job at hand. Where plumbing codes permit, this usually means using plastic pipe, which is quickly cut with a saw and easily spliced with cement. If you must use copper tubing, see if the flexible type will do: it bends easily around corners, saving time and money by reducing the need for fittings. Since it can be joined with flare or compression fittings *(pages 73-74)*, flexible tubing is also ideal for situations where soldering joints with a blowtorch is impractical or dangerous.

Whatever pipe material you choose, the job will go more easily if you follow a procedure used by professional plumbers: complete as much of the assembly as possible at your workbench. Even longer runs of pipe can be partially fabricated there, section by section, then brought to the place where they are to be installed. This method allows you to do much of the task in a well-lit, comfortable work area instead of the dark, cramped places where pipes must sometimes be run. If you prefer soldered joints, it also reduces the amount of soldering that must be done near the flammable structure of the house and its wiring.

Putting in a new fixture often necessitates drilling small holes in a wall to locate the existing pipes, then cutting a larger hole to allow enough work space to cut into these pipes in order to start the new branch lines. This need not deter a would-be plumber, since holes a foot square or even larger are generally easy to patch with wallboard, plaster and joint tape, since none of these holes are made in the structural, load-bearing beams of the house. But to keep wall cutting to a minimum, you should run new pipes along the outside of a wall wherever you can, and then create a camouflage for them with framing or cabinets.

A Range of Materials and Fittings for Every Task

The word plumbing comes from the Latin for lead, and there are still a fair number of lead pipes and fittings around. But today a variety of materials is used in house plumbing—copper, cast iron, steel, brass and plastic. Several are likely to be combined in a house, for each brings a balance of benefits in economy, durability and convenience that makes it the choice for a particular job. And the combination cannot be willy-nilly, for some materials cannot join others without special precautions.

Copper, most popular for supply lines but also used for drain and vent lines, is convenient to work with and very durable but costly. It is available as rigid pipe in 10- and 20-foot lengths or as flexible tubing in 30-, 60- or 100-foot coils. Rigid pipe is tougher, but flexible tubing is easier to install, particularly when adding to existing plumbing. Rigid pipe must be soldered with sweat fittings, while flexible tubing can also be joined mechanically with flare or compression fittings. Flexible tubing comes in two weights: Type K, the heaviest, mainly for underground lines outdoors, and medium-weight Type L. Rigid pipe comes in Types K and L, and in lightweight Type M, generally adequate for homes.

Plastic pipe has become increasingly popular for drains and, in some areas, supply lines because it is light ($^1/_{20}$ the weight of the same size steel pipe), inexpensive, easy to join and seldom bursts even if the water inside freezes. Several different plastics are made into pipe, but only CPVC and polybutylene can be used for hot-water supply lines. Other plastics, such as ABS, can be used for drain and vent lines. Like copper tubing, plastic pipe is available in rigid or flexible form. The rigid pipe is joined with solvent and glue. Flexible pipe—either polybutylene, suitable for hot and cold lines, or polyethylene, suitable for cold water—is joined with insert fittings and metal clamps and is usually used only for outdoor supply lines.

Galvanized steel is the strongest material available for water supply lines and is preferred for piping exposed to damage. It is heavy, and must be joined with threaded fittings. Most plumbing-supply stores will cut the lengths you need and thread them.

Cast-iron pipe, the heaviest and most durable of all piping materials, is used only for drains and vents. Today it is joined with metal-clamped rubber gaskets called hubless fittings—old pipes have special lead-caulked fittings.

The practical considerations of convenience, cost and durability, however, are not the only factors determining the choice of material. Your local plumbing code forces decisions. Some codes prohibit a material in one part of a plumbing system but not in another, and even dictate the method you may use to join it. Check with your building department before buying any materials.

All piping is sized by inside diameter. When replacing a section, measure with a ruler its inner diameter and get new pipe the same size. If you are adding a supply line, use the pipe size specified by the manufacturer of the fixture. Choose a drain pipe size by using the guidelines on page 85.

In addition to piping, you also need fittings to join it. A fitting is necessary whenever piping branches off, changes diameter, joins another type, or in the case of rigid pipe, whenever direction changes. The most common fittings are pictured opposite. The shapes are common to all piping—the only difference is the way they are joined.

A special precaution must be taken when joining copper pipe to steel: wrap the threads of the copper fitting with plastic joint tape to prevent an electrochemical reaction that occurs between dissimilar metals, eroding the joint.

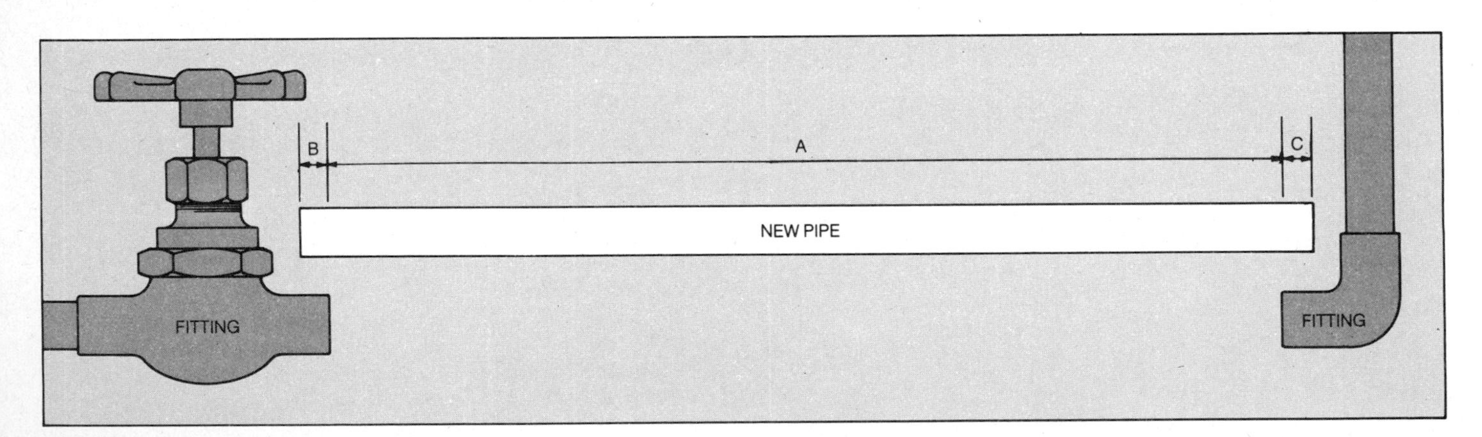

How much pipe do you need? To find out how long a piece of pipe you need between two existing fittings, measure the distance between the faces of the two *(distance* A). Then measure the distance that the new pipe will extend into each fitting *(distances* B *and* C), and add both to the first. If only one fitting is in place, mark where the second fitting will be placed and have someone hold it there while you measure.

Four Categories of Connectors

Branches and turns. An elbow takes rigid piping around a curve; various angles are available up to 90°. A T provides a connection for a new branch at its base. A Y is a variation of the T usually found in drainage and vent lines to allow pipes from individual fixtures to be joined to one main stack. The branch of a Y can also be plugged for use as a cleanout.

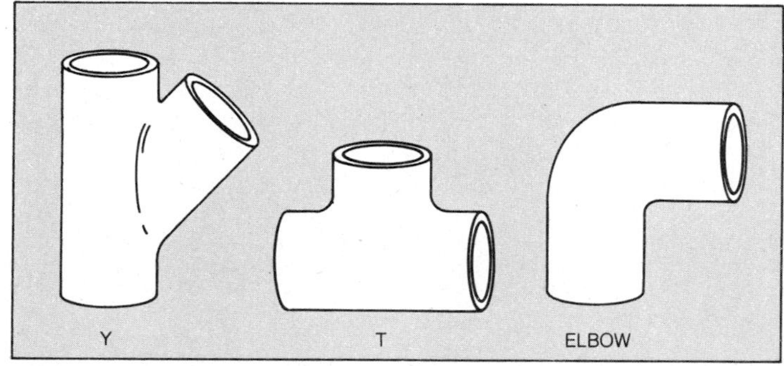

Fittings for in-line joints. When two pipes must be connected together in a straight run, a coupling, or sleeve, splices them. A nipple threaded on both ends extends a coupling for steel pipes. Ready-made nipples come in graduated sizes up to 12 inches long. A reducer attaches a length of small-diameter pipe to a larger one. Ts, elbows and Ys are available with a reducer.

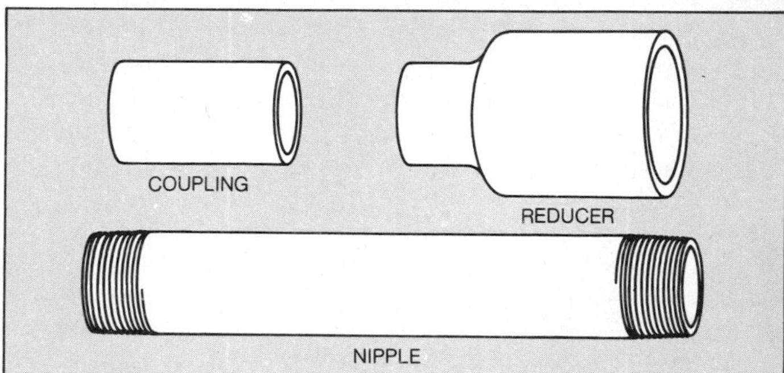

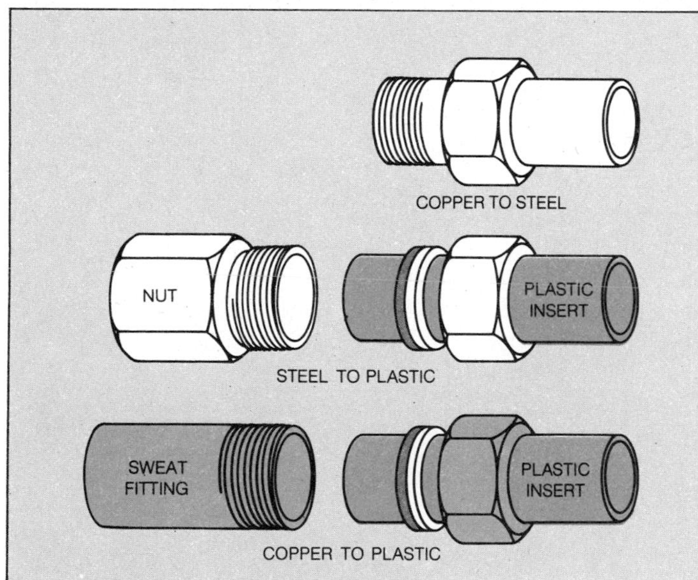

Transition fittings. To connect one type of pipe material to another, a special fitting is needed to fasten both materials firmly and prevent the electrolytic action that can cause corrosion at joints between dissimilar metals. The copper-steel fitting has one threaded end, to fit steel piping, and one smooth end to solder to copper pipe. The plastic-steel fitting has one threaded steel nut and a plastic insert to which a length of plastic pipe is glued. Because the plastic is electrically inert, the two materials need not be separated. The plastic-copper fitting is similar but with a copper sweat fitting. Transition fittings are also available to join copper, steel and plastic drain pipe to cast-iron pipe.

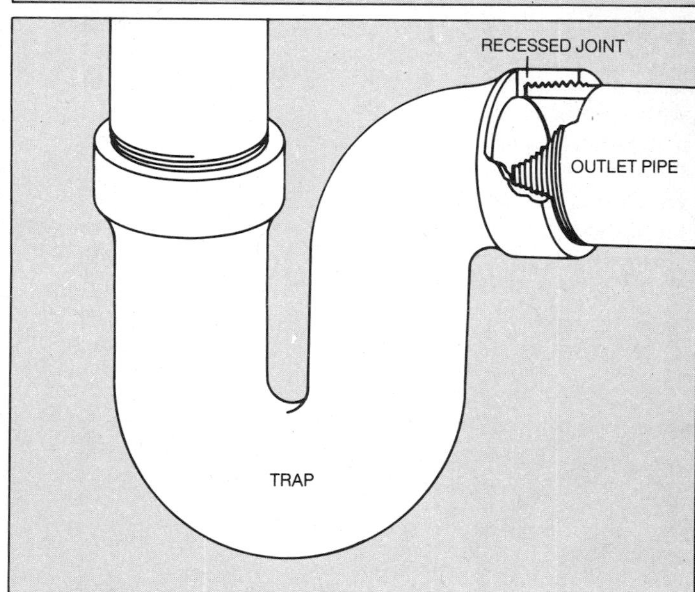

Drainage fittings. The trap, the one fitting used only in the drainage system, keeps a barrier of water in its U-turn to seal out odors from sewers. All drainage fittings, whether threaded or not, have recessed joint surfaces so that when a pipe is inserted the inner surface of both pipe and fitting will be flush, eliminating any possibility of waste material snagging in the joint and causing a blockage. Since drain lines depend on gravity to pull waste through the system, the outlets of drainage fittings angle downward.

Copper Piping

The techniques for joining copper piping and fittings are the same whether you are adding new runs or repairing damage in old ones. The first essential step is to cut the pipe or tube perfectly straight, then clean its end of kinks or burrs and clean the mating surfaces of both pipe and fitting. Flexible and rigid copper piping can be cut with a special tool called a tube cutter, or with a hacksaw and miter box. For flexible tube, you may need a bender. A well-made soldered sweat joint is the best for either rigid pipe or flexible tube, stronger and more leak resistant than a mechanical connection. When soldering, use noncorrosive flux and 50-50 tin-lead solder, not solder with an acid or rosin core.

Caution: Before working on plumbing with a torch, drain all water from the pipes to avoid a dangerous steam buildup. Place asbestos board between the pipe and surfaces that might ignite.

Because of the fire hazard, avoid soldering in cramped spaces; instead of sweated joints, use flexible tube and mechanical connectors if you can. These compression, tap-on and flare fittings go only with flexible tube; they are somewhat more expensive and less durable than sweat fittings, but they require little skill to make and are generally preferred for small jobs.

Getting the Right Length and the Right Shape

Using a cutter. Slide the cutter onto the pipe or tube and turn the knob until the cutting wheel bites into the wall. Do not tighten the knob all the way or it may bend the wall and the joint will leak. Turn the cutter once around, retighten the knob, and continue turning and tightening until the piping is severed. Use the triangular blade attached to the cutter to ream out the burr inside and, with a file, remove the ridge that the cutter has left on the outer surface. If you cut with a hacksaw, remove the inner burr with a round file.

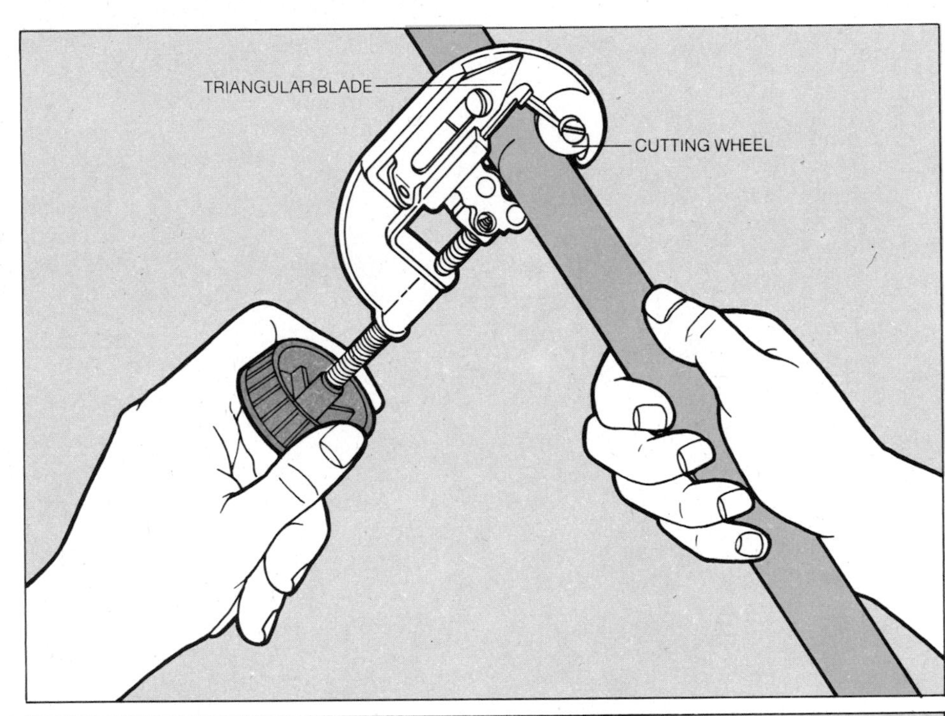

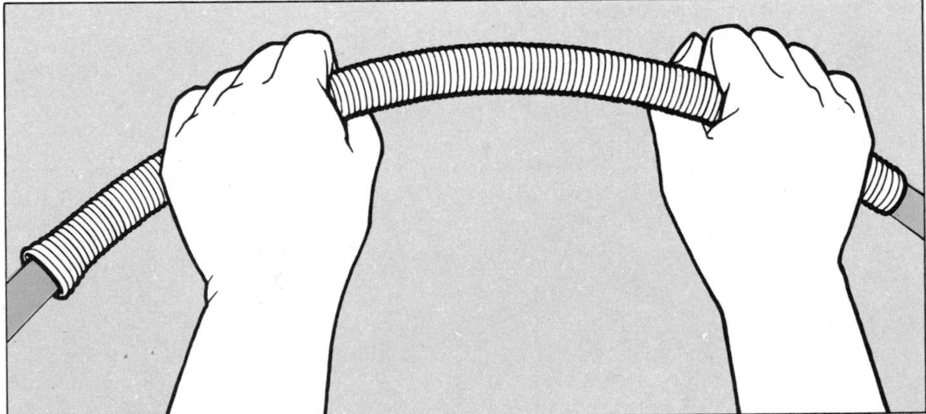

Using a tubing bender. Unlike rigid copper pipe, flexible tube can be bent for turns. To prevent kinks in the walls, slip a coiled-spring bender over the section, using a clockwise twisting motion. Bend the tube with your hands or form it over your knee. Overbend the tube a bit, then ease it back to the exact angle you want.

Tap-on Fittings: The Simplest Way to Tap a Line

1 Tapping the existing line. Remove the valve from the tap-on fitting and clamp the yoke around a ½-inch supply line, which can be either copper, brass or galvanized steel pipe. Drain the water from the line and drill a hole through the wall of the pipe at the center hole of the yoke. Be careful not to drill through the other side.

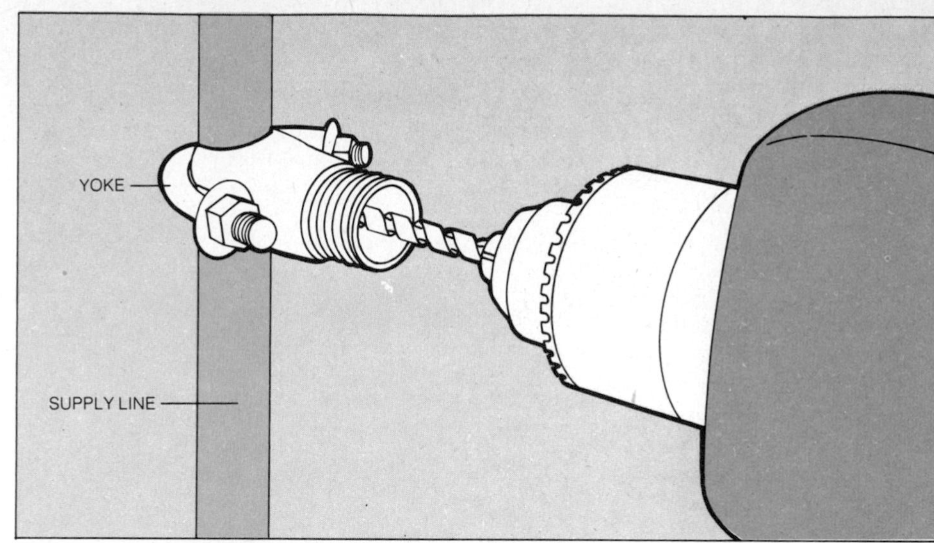

2 Installing the valve. Thread the valve back into the center hole of the yoke. Make sure that the yoke is clamped securely to the pipe and connect flexible tube to the valve with a compression fitting. The fitting shown accepts ½-inch tube —large enough for most fixtures—or a faucet. A similar but smaller fitting accepts ¼-inch tube, used mainly to supply small appliances like humidifiers and ice-cube makers.

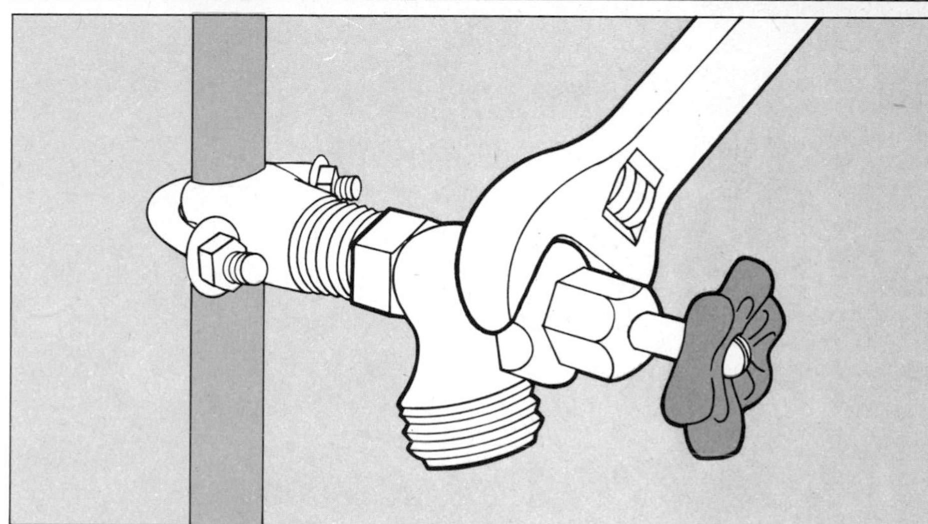

Compression Fittings for Flexible Tubes

Assembling a compression fitting. Cut the supply-line tube square; remove burrs, inside and out. Slip first the flange nut, then the compression ring onto the tube and insert the tube into the fitting—a cutoff valve for a fixture in the drawing—as far as it will go. Slide the compression ring into the joint, making sure that it is squarely aligned, then slide the flange nut over the fitting's threads and screw it down until it is hand-tight. With a wrench, tighten the nut another quarter turn but no more: overtightening will damage the compression ring and cause a leak.

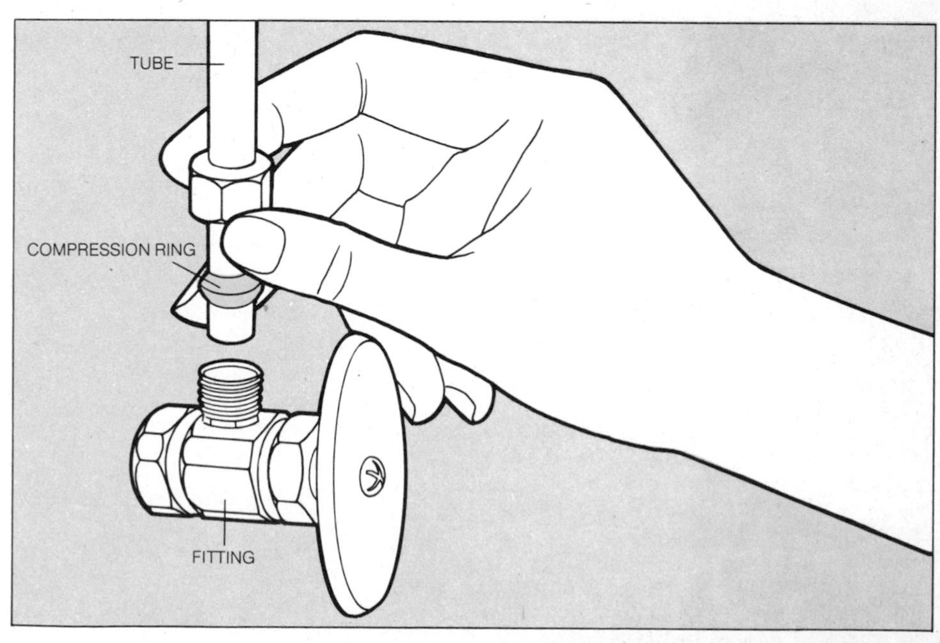

Making a Flare Joint

1 Flaring the tube. Slip a flare nut over the end of square-cut deburred tube and insert the tube into the hole sized for it in a flaring die. Bring the end of the tube flush with the face of the die and tighten the wing nuts. Position the flaring tool over the end of the tube. Clean the point of the tool, then turn the handle clockwise until the point enters the tube. Make sure that the point is centered, then continue tightening until the tube end is flared to a 45° angle. Retract the tool by turning the handle counterclockwise.

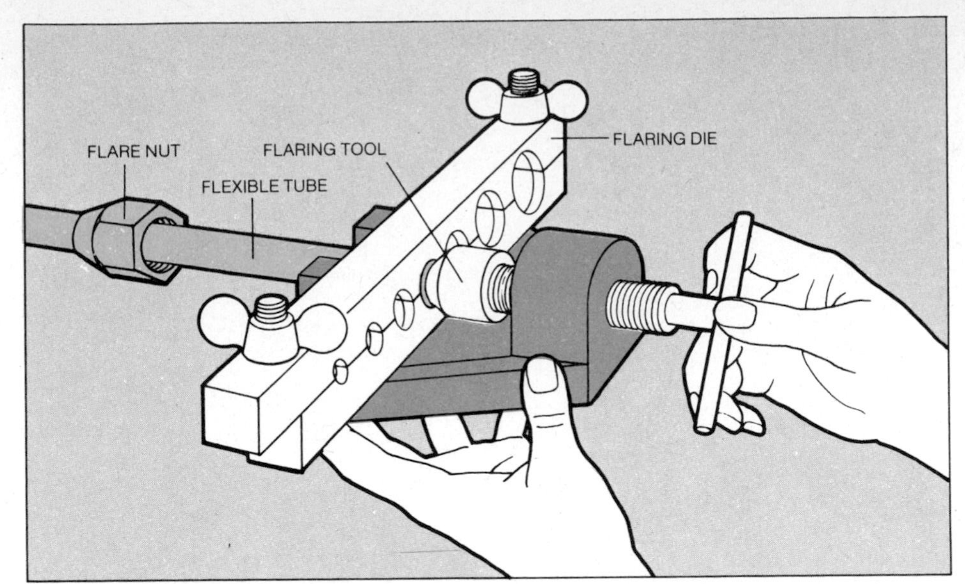

2 Assembling the joint. Set the flared end of the tube onto the domed end of the fitting—a straight connector is illustrated but elbows, Ts and other types are also made. Then slide the flare nut up and thread it to the fitting until it is hand-tight. Finish tightening the nut with a pair of open-end wrenches, one on the flare nut and one on the fitting. The other end of the flare fitting is joined to a second piece of tube in the same way.

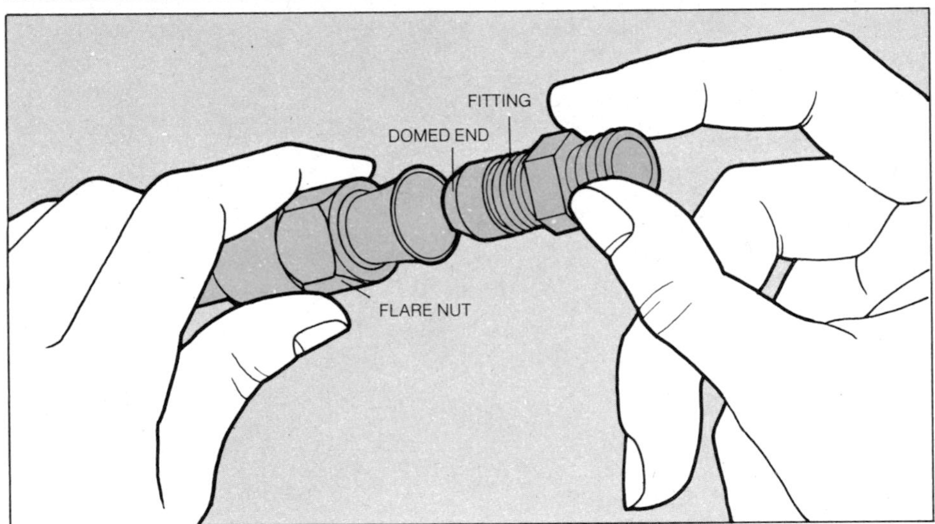

How to Use a Propane Torch

The small propane torches generally used to sweat copper joints are made in two parts, a combination valve and nozzle assembly, and a replaceable metal tank of fuel. To assemble the torch, simply screw the nozzle assembly to the threaded fitting at the top of the tank.

To light the torch, strike a match, hold it near the nozzle and turn the valve slightly clockwise. When the torch lights, gradually open the valve further until the flame becomes large enough to heat the area you will work on. Do not open the valve all the way: the flame will not get much larger and the gas pressure will probably blow it out.

It is necessary to hold the torch essentially upright while you are working. If the tank is tilted very much, the liquid propane inside will flow into the valve, blocking it so that the flame goes out.

The propane torch is easy and safe to use if it is handled properly. Caution is necessary, however, because the joints to be sweated with the flame are generally located near flammable parts of a house structure.

Before starting to work, cover the area behind piping with squares of asbestos sheeting. And make a habit of shutting off the torch whenever you set it aside, even if only for a moment or two—the flame is silent, nearly invisible and easy to forget about.

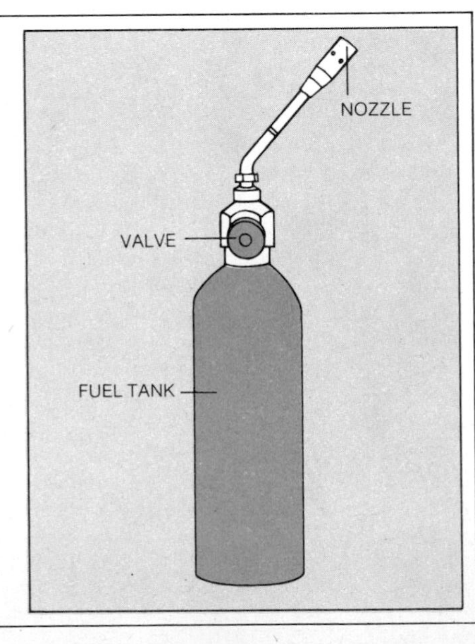

Sweating a Copper Joint

1 Cleaning the copper. To clean the joint so that melted solder will flow evenly and adhere securely, scour the inner surfaces of fitting sockets with a wire brush. With a piece of emery cloth—not a file or steel wool—clean the end of the pipe that will slide into the fitting socket, rubbing until the surface is burnished bright. Once surfaces are cleaned, do not touch them, since even a fingerprint will weaken the joint.

2 Assembling and heating the joint. Brush a light coating of flux over the surfaces you have cleaned, assemble the joint and give the pipe a twist to make sure the flux is distributed evenly. Place a piece of asbestos in back of the joint to protect the surface behind from flame. Light the torch and play the flame over the fitting and nearby pipe, heating them as evenly as possible. Touch a piece of solder to the fitting and then to the pipe: when the solder melts on contact with both parts, the joint is ready to be soldered. Do not heat further or the flux will burn off and the solder will not flow properly.

3 Soldering the joint. Touch the solder tip to the point where the pipe enters the fitting, but do not let the torch flame touch the solder. The solder should melt only on contact with the hot metal; if it does not, take the solder away and continue heating the joint. (Experienced plumbers judge the temperature of a joint so accurately they heat it and remove the torch flame before soldering.) When the joint is properly heated, the flux inside draws molten solder into the fitting to seal the connection. Continue feeding solder to the joint until a bead of metal appears around the rim and begins to drip.

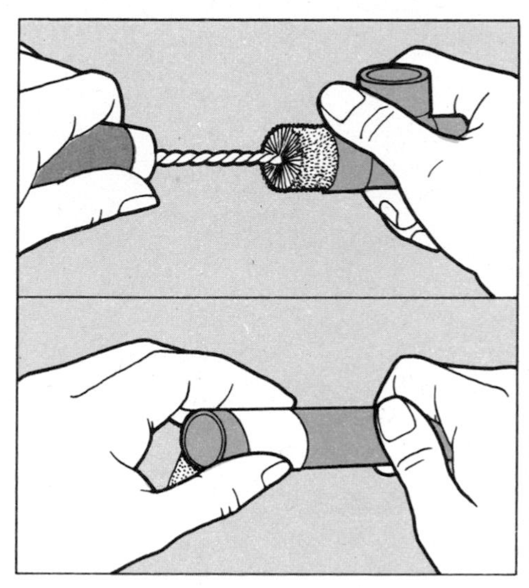

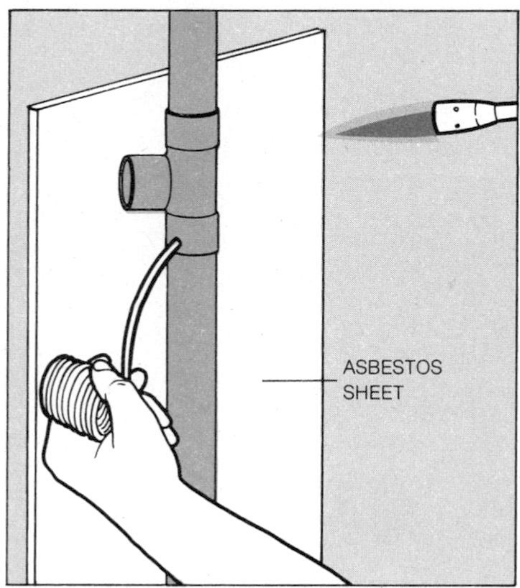

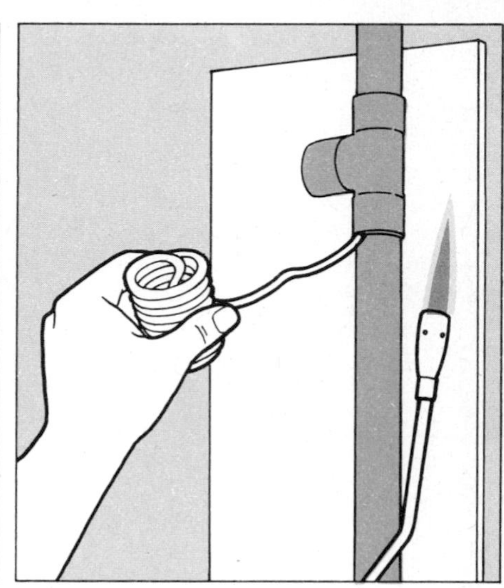

Stopping Leaks in Copper Pipe

Fixing a leaking joint. Leaks in sweat joints are almost always caused by a poor soldering job. Drain the water from the line, protect adjacent surfaces with asbestos and heat the joint until it can be pulled apart. Clean the mating surfaces of both pipe and fitting, then apply flux as described for a new joint. Heat pipe and fitting separately, and cover the fluxed surfaces with solder. When both surfaces have been "tinned," scour them smooth with emery cloth, assemble the joint, and heat and solder it as described for a new joint.

Compression and flare joints leak either because of poorly cut pipe or improper assembly. Disassemble the joint by loosening the nut, and check to see if the nut and fitting have been cross-threaded. If they have, the entire joint must be replaced. In a compression fitting, also check that the pipe end and compression nut are not bent out of shape. Replace a misshapen compression nut and cut off a distorted pipe end. For a flare joint, check the inside of the flared pipe for burrs, gouges or distortions in the flare. To cure any of these problems, cut off the damaged end perfectly square and make a new flare.

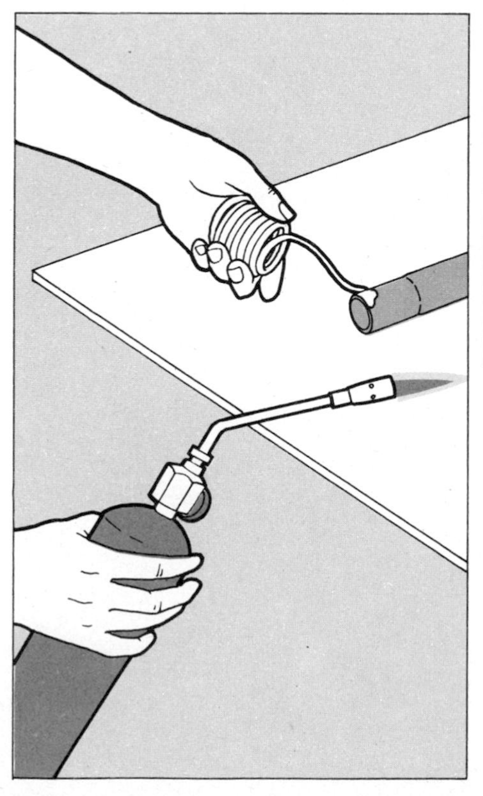

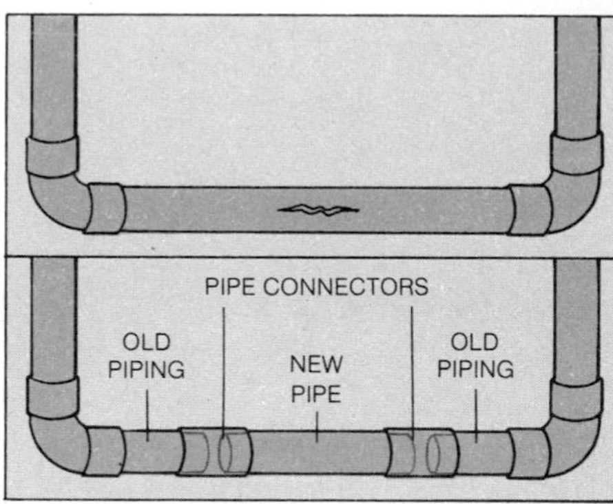

Replacing damaged piping. If a section of copper piping is punctured or burst, it can be patched temporarily but must eventually be replaced. Drain the water from the line and cut out the damaged section *(top)*. Remove only enough pipe to leave undamaged piping with square-cut ends. Replace the section with new piping attached by straight connectors at each end—sweated joints for rigid pipe, either sweated or mechanical ones for tube.

Plastic Piping

One of the main reasons that plastic pipe has become so popular is the ease with which it can be assembled. Rigid pipe and fittings are simply cemented together, and the resulting joint is stronger than the pipe itself. Flexible plastic tubing can be joined with insert fittings or compression fittings like those used with copper tube. Transition fittings *(page 71)* make it easy to start a run of plastic pipe from existing pipe of a common material.

Before buying plastic pipe, check your local code: some codes ban it, while others permit it to be used only outdoors, only for drains or only for cold-water supplies. If you use plastic for hot-water lines, reset the temperature-pressure relief valve on your water heater for no more than 180°, the maximum safe operating temperature for plastic pipe. Even at lower temperatures, rigid plastic pipe carrying hot water becomes slightly flexible and it should be supported with clamps placed every 3 feet and set loose enough to let the pipe move slightly.

Assemble rigid plastic pipe only when the air temperature is above 40°; cold slows the action of the solvent-cement, interfering with bonding. Caution: Cement vapors are hazardous. Work in a well-ventilated area.

Cemented Fittings for Rigid Pipe

1 Cutting the pipe. Flexible plastic pipe is soft enough to be cut with a sharp knife or even a large pair of shears, but cutting rigid pipe demands precision. A tube cutter that is used for flexible copper tube is convenient, but it needs a blade meant for plastic. An equally accurate cut can be made with a miter box and backsaw or a hacksaw with a blade having 24 teeth per inch.

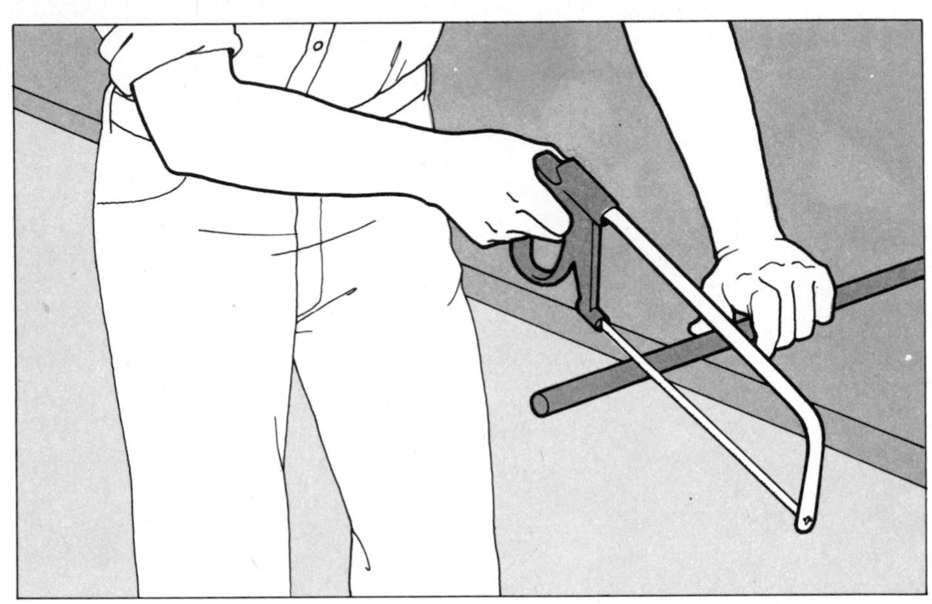

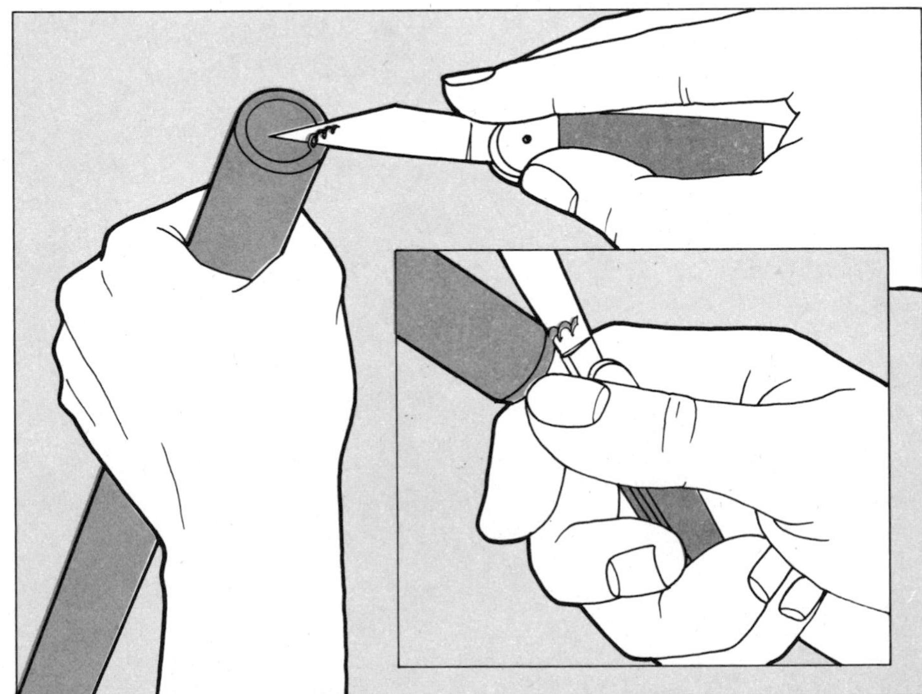

2 Preparing the pipe end. Pare away any burrs from the inside of the pipe end. Then bevel the outside of the end so it will not force the glue from the inside of the fitting *(inset)*.

3 **Checking the fit.** Clean the pipe end and fitting with a dry rag to remove any grease or moisture, then slip the pipe into the fitting: you should be able to push it in about halfway. If the fit is too tight, the glue may be forced from the joint; too loose and the pipe and fitting may not bond together. In either case the joint will leak: discard the fitting and try another (size varies slightly). When the fit is correct, adjust pipe and fitting to the position in which they will be set. Mark them so they can be quickly repositioned after glue has been applied.

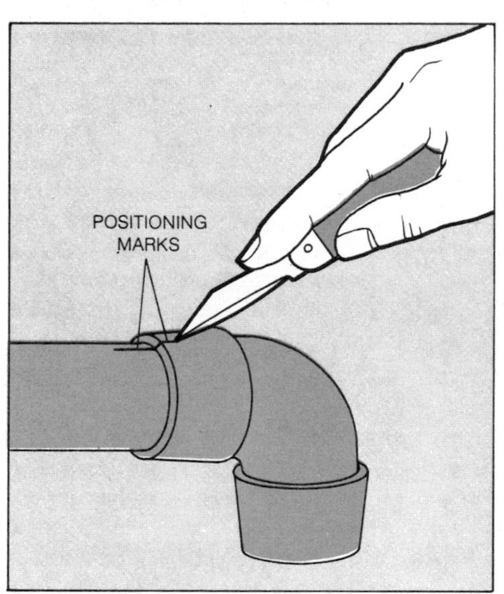

4 **Priming and gluing the joint.** Although cement alone generally makes a satisfactory joint, many plumbers prefer to clean and prepare the mating surfaces of the pipe and joint by brushing on a coat of priming solvent. After applying the solvent, wait about 15 seconds to allow the surfaces to soften. While the primer is still wet, use a wide brush to apply a thick coat of cement to the primed surface of the pipe and a thinner coat to the inner surface of the joint. Do not apply so much cement that it blocks the pipe opening. Immediately slip the pipe into the fitting, twist the pipe a quarter turn to evenly spread the glue, and align the pipe and fitting to the marks made in Step 3. Delay may cause the joint to fuse in the wrong position. An even bead of glue should appear around the joint. If the bead is incomplete *(below)*, pull the pipe out immediately and apply more glue. When the joint is properly set, hold it in place for 30 seconds, then wait three minutes before starting the next joint. When the job is done, wait at least one hour—or, better still, overnight—before letting water into the new run of pipe.

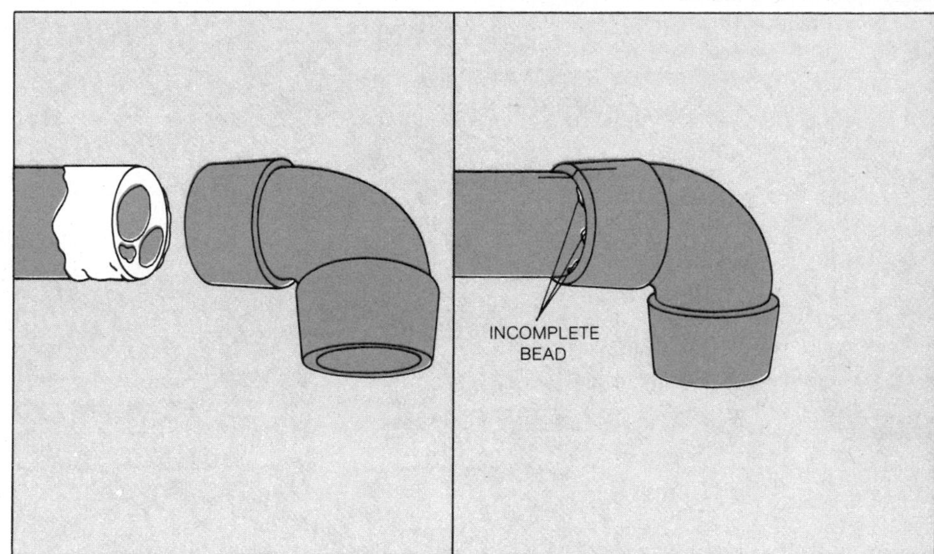

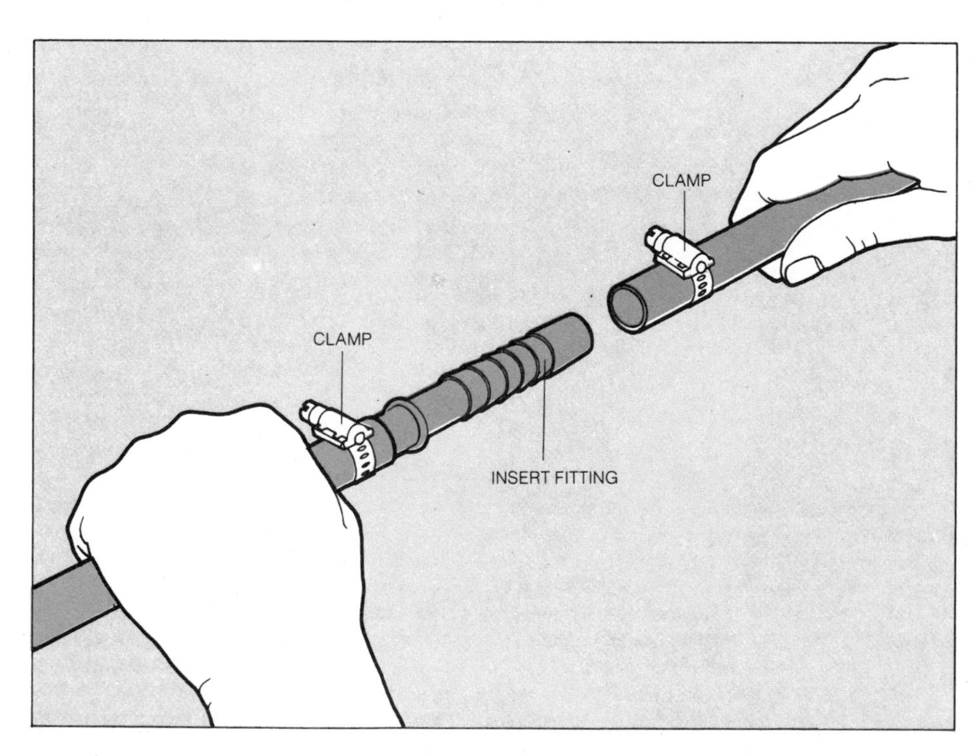

A Clamped Fitting for Flexible Tubing

Attaching an insert fitting. Flexible plastic tubing —polyethylene for cold water and polybutylene for hot—is generally joined by a ridged insert fitting and two small clamps. Slip a clamp loosely to about ½ inch from the end of each pipe. Force the pipe ends over the fitting until the ridges are covered. Tighten the clamps with a screwdriver.

Steel and Brass Piping

Galvanized steel pipe and fittings, as well as the less common brass pipe often found in older homes, must be joined with threaded joints. If you plan only a small extension or replacement job, take careful measurements *(page 70)* and have the plumbing-supply store cut and thread the pipe for you. For a more extensive job, use copper or plastic for it, making connections with adapters.

Threaded joints make assembly easy, but complicate the job of removing a damaged pipe or inserting a new fitting in existing pipe to make a new branch *(page 87)*. The problem is that once a pipe or fitting is in the line it cannot be unscrewed as one piece, since loosening it at one end will tighten it at the other. The solution is to cut out the old pipe and install a new one using a fitting called a union *(opposite, top)*. A simpler device called a slip fitting—a sleeve sealed at both ends by rubber gasket-fitted compression nuts—*(opposite, bottom)* can be used to fix a leaking pipe, and tap-on fittings like the one on page 73 are also available for steel pipe.

Assembling Threaded Pipe

Applying joint sealers. Before assembling a joint, pipe threads should be covered with one of the materials below to lubricate, seal, rustproof and, if necessary, allow easy disassembly. Pipe-joint compound *(left)*, the most common material, is spread evenly over the pipe end; use just enough to fill the threads. If you wish to make doubly sure that the joint will not leak, wind wicking into the pipe threads *(center)* before applying joint compound. Use both joint compound and wicking before reassembling an old joint. Plastic joint tape *(right)* is even easier to apply. Wind one and one half turns of tape clockwise over the threads tightly enough for them to show through.

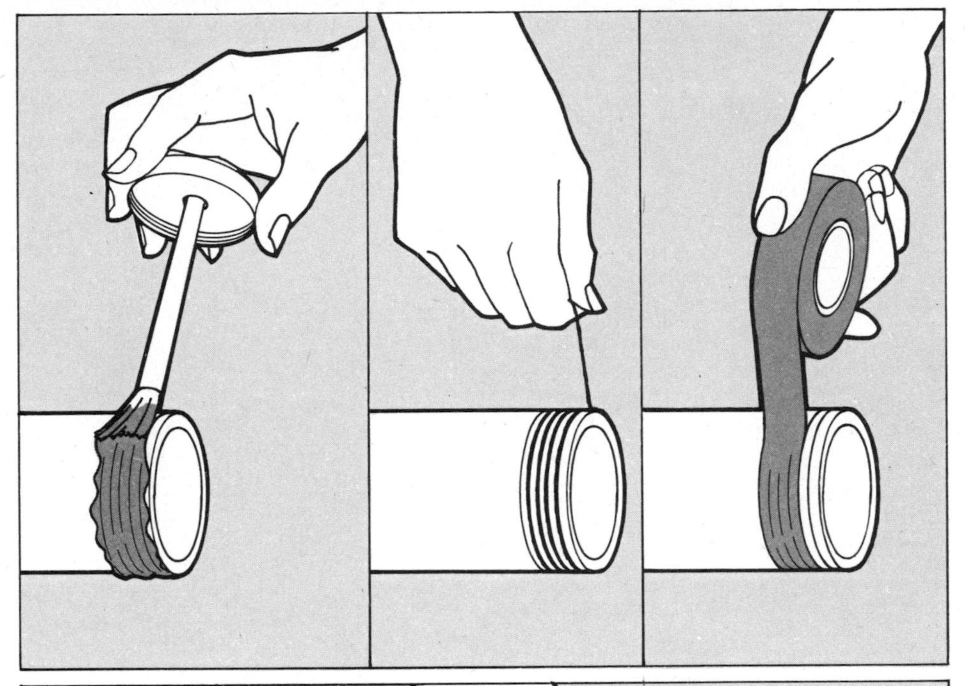

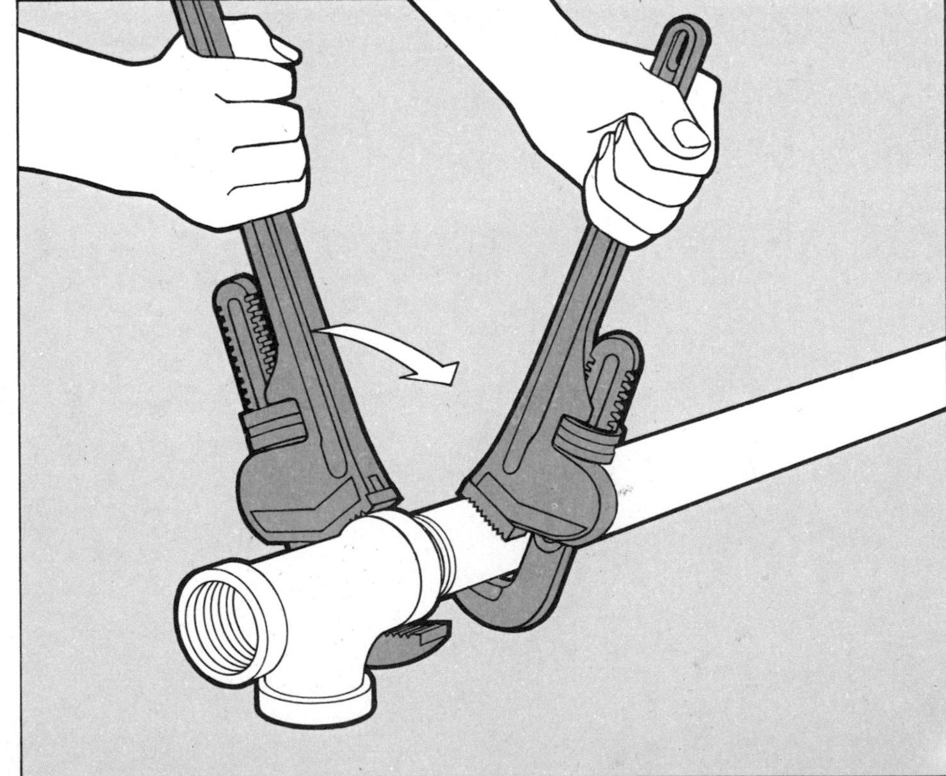

Joining the pipe and fitting. Thread pipe and fitting together by hand to ensure that they are not cross-threaded, then use two pipe wrenches to finish tightening the joint. Turn the fitting with one wrench and hold the pipe steady with the other so that the rest of the pipe branch will not be twisted or strained. The jaws of the wrenches should face the direction in which the force is applied *(arrow)*. Keep tightening until just three pipe threads are visible outside the fitting. Further tightening may strip threads and cause a leak.

Replacing Damaged Pipe with New Pipe and a Union

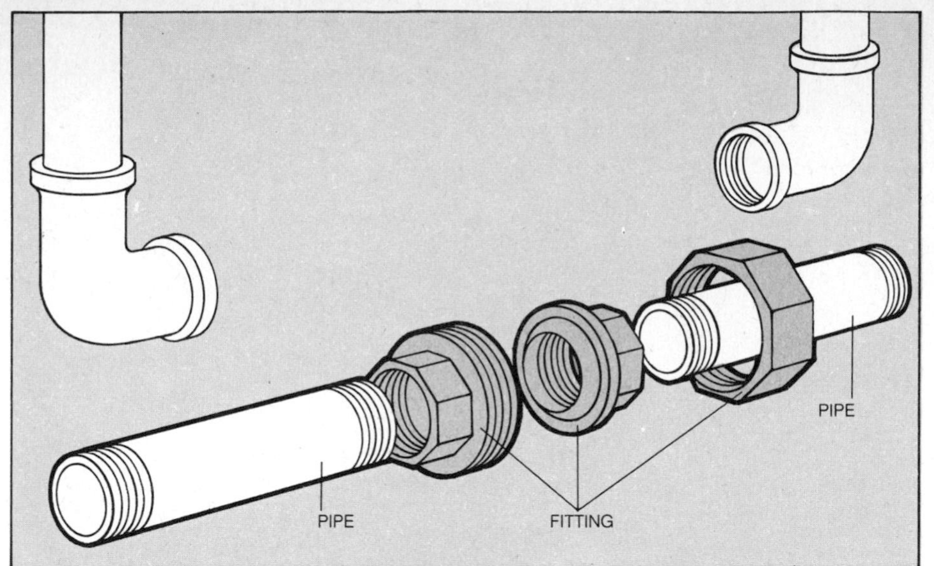

1 Preparing the parts. Hacksaw through the damaged section and unscrew each piece of pipe from its fitting by reversing the procedure shown opposite, bottom. Buy two pieces of pipe and a union whose combined length when assembled will be the same as the old pipe. Prepare the threads of one new piece, slip one union nut onto it and tighten. Slip the ring nut over the end of the second pipe, then thread the other union nut onto it and tighten it.

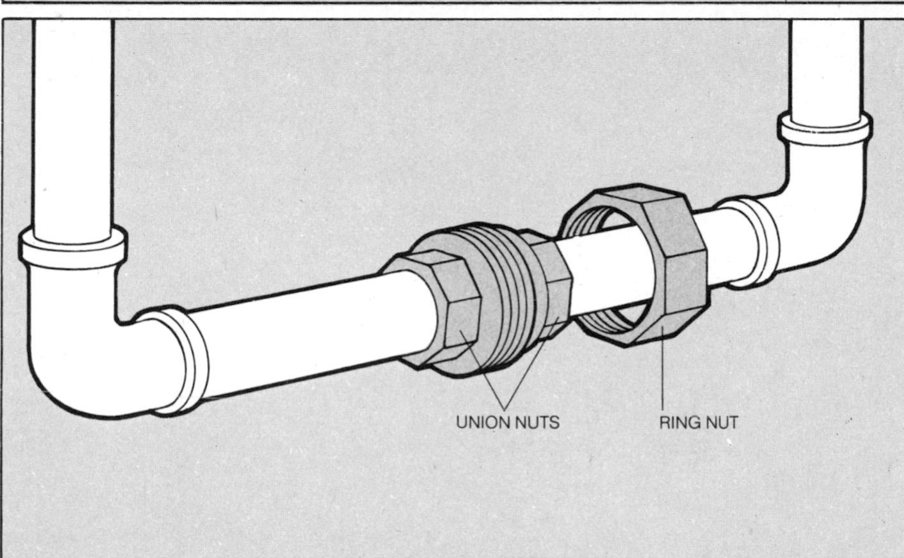

2 Connecting the union. Prepare the threads at the end of each pipe section opposite the union nut and screw it into one of the existing fittings. When both sections are tightened, the faces of the union nuts should touch. Slide the ring nut to the center of the union and screw it on the exposed threads of the union nuts. Tighten it in place with one wrench while bracing the exposed union nut with another wrench.

A Sleeve to Stop a Leak

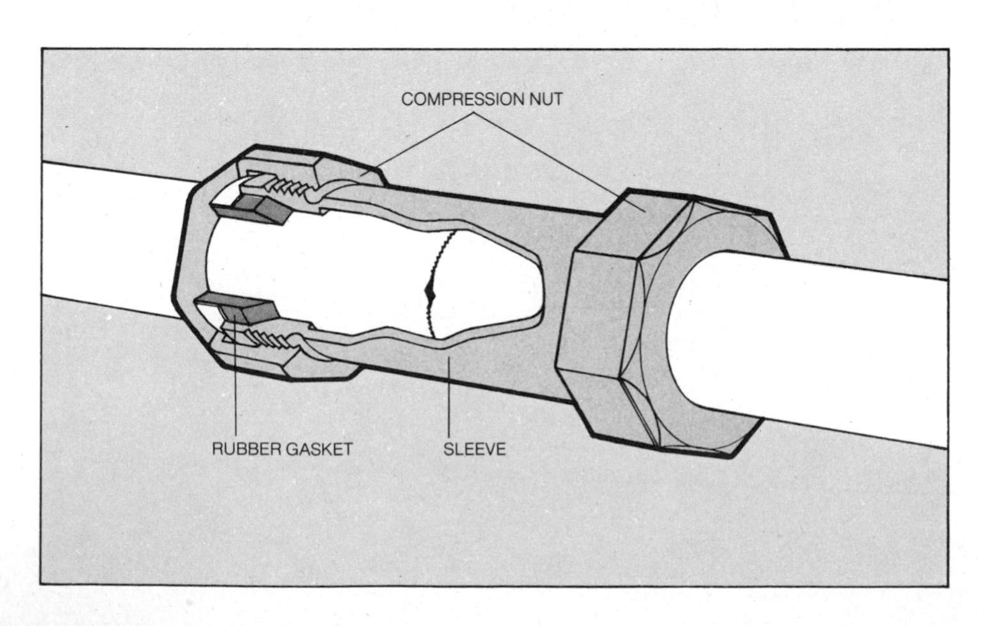

Installing a slip fitting. Cut through the damaged pipe with a hacksaw, pull one cut section aside and slide the fitting onto it. Realign the two sections and slide the fitting over the damaged area. Tighten the nut at each end of the fitting. Since this fitting is not as strong as a union, it should be used only on exposed pipe where you can tell if the leak starts up again.

Cast-Iron Piping: Old and New

Because of its low cost and durability, cast iron has long been the most popular material for drain, waste and vent pipe, but until recently it was also the most difficult to install because of its weight and the difficulty of joining it. The conventional type, still in wide use, is connected at bell-shaped hub joints *(page 19)*, which must be sealed with molten lead and a ropelike material called oakum. A newer type, called hubless pipe, is joined with sleeves and clamps *(opposite)*. It can be spliced into an existing system of hub-type pipe to replace a damaged section or to add a new one *(page 84)*. Leaks at hub joints can often be stopped by using the method shown on page 19.

Two grades of cast-iron pipe are available, service weight and the thicker-walled heavy weight. Use service weight if your plumbing code permits, because it is easier to work with. Both weights can be cut either with a saw and chisel *(below right)* or with a special cutting tool *(below left)*. Proper support is more important with cast-iron pipe than any other type, both because of its weight and because the hubless joints are slightly flexible. Use pipe strap or the special clamps shown opposite, bottom.

Joining Hubless Pipe

1 **Cutting the pipe.** The easiest way to cut cast-iron pipe is with the cutting tool shown at left below, usually available at rental stores. If you are working inside a wall *(page 86)*, there may not be enough space to cut the pipe any other way. Wrap the chain section of the tool around the pipe, hook it onto the body of the tool, tighten the knob and work the handle back and forth until the pipe snaps.

Cast-iron pipe can also be cut with hacksaw and chisel *(below right)*. Mark the circumference of the pipe where you wish to sever it, then support the pipe on a 2-by-4 so that the section to be cut off is raised. With a hacksaw, cut a 1/16-inch-deep groove around the pipe at the mark, then tap around the groove with a hammer and chisel until the pieces separate.

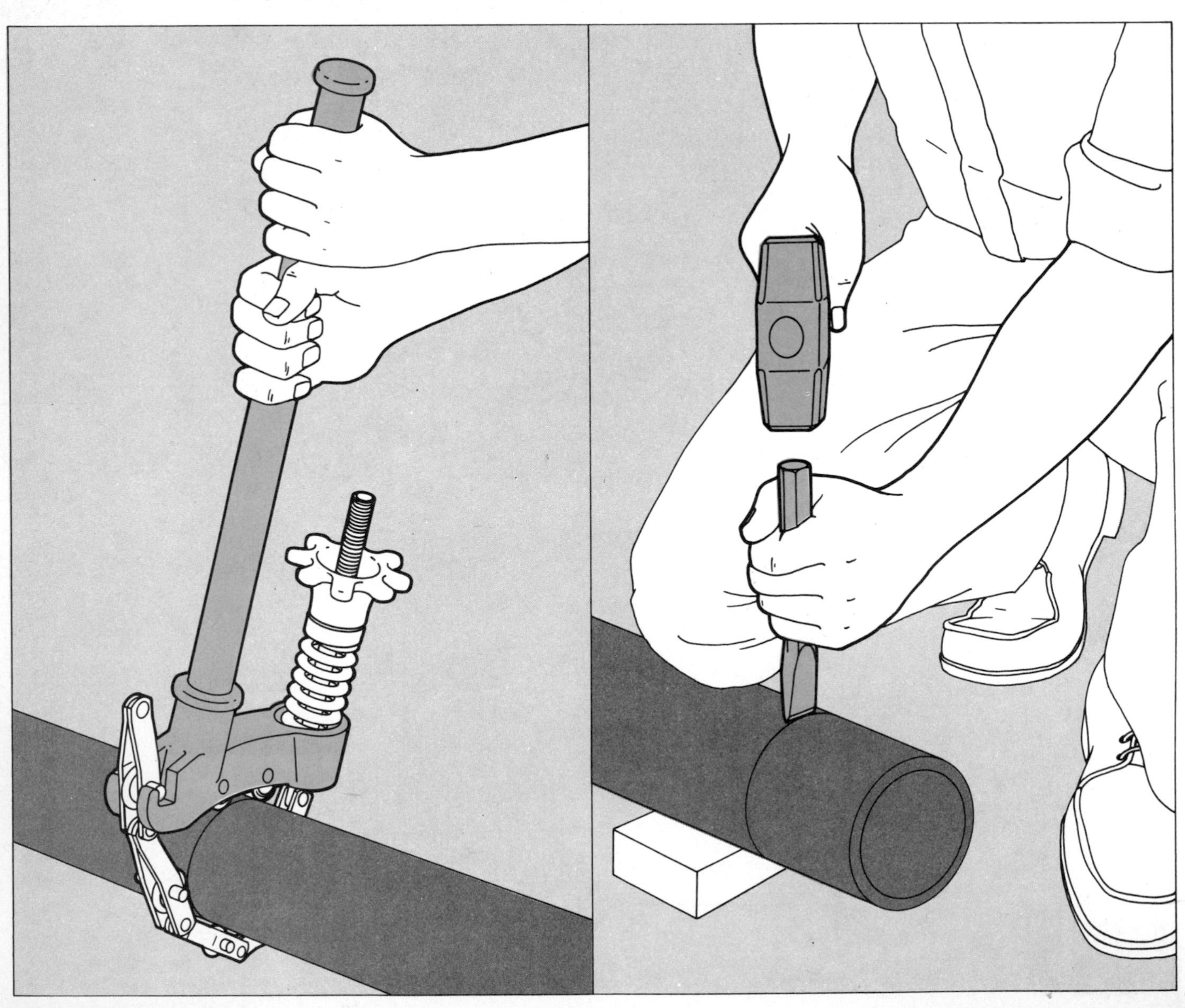

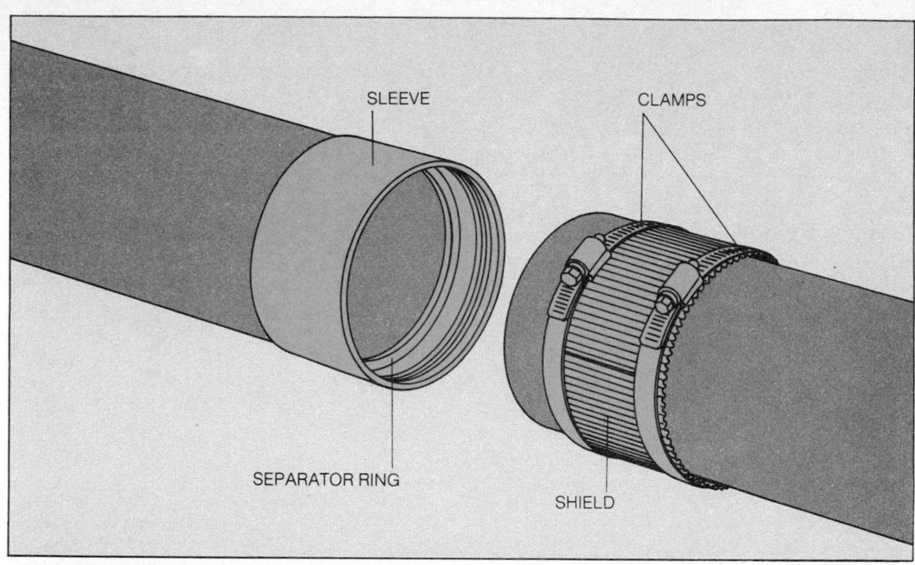

2 Preparing the joint. Slip the rubber sleeve onto the end of one pipe, making sure that the pipe end butts firmly against the separator ring at the center of the sleeve. Slide the stainless-steel shield over the end of the second pipe, and position it so that the clamps will be accessible for tightening when the joint is assembled.

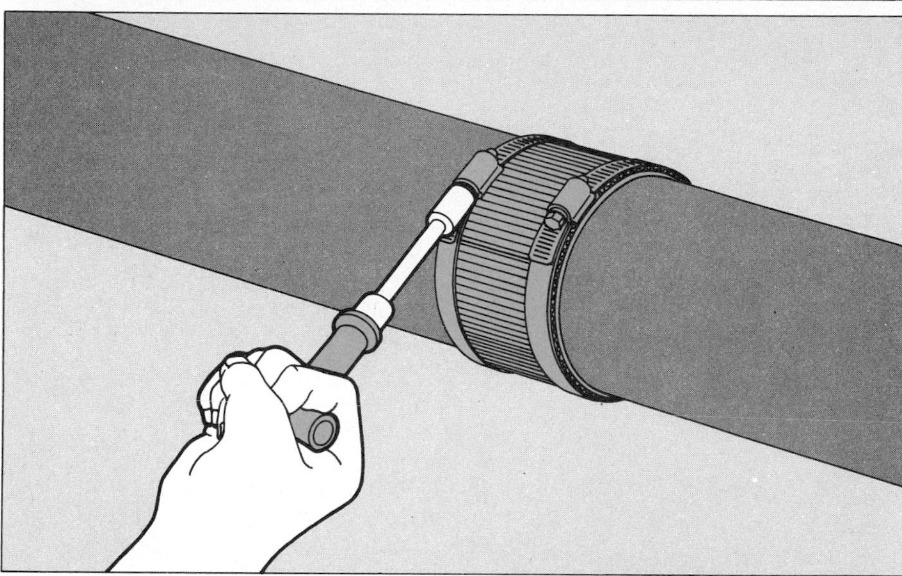

3 Assembling the joint. Push the end of the second pipe into the sleeve until it butts against the sleeve separator ring. Slide the shield over the sleeve so that it covers the sleeve completely, then tighten the clamps on the shield. To tighten the clamps, you may need either a screwdriver or the special wrench shown in the drawing.

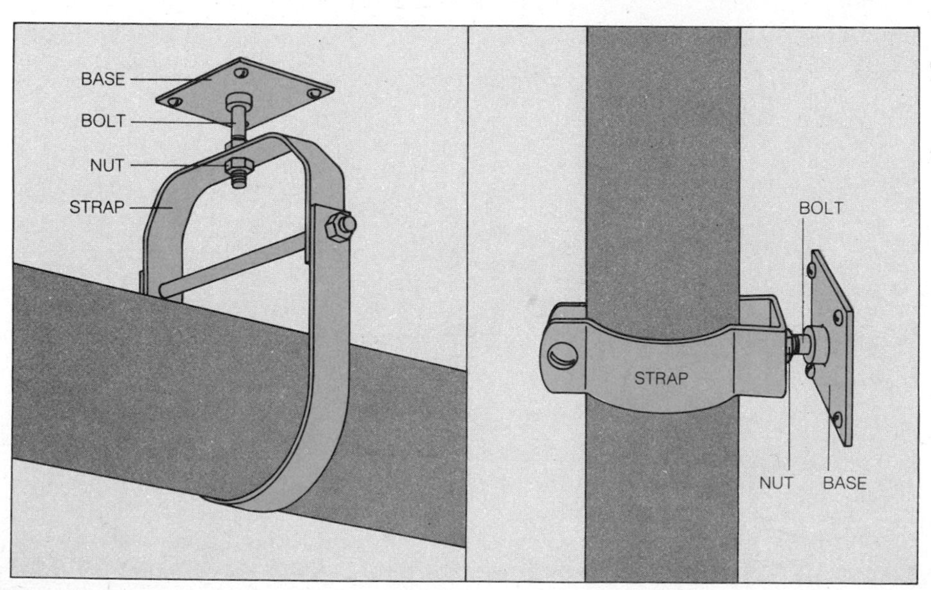

4 Supporting the pipe. A horizontal run of hubless cast-iron pipe should be supported at every joint with a special clamp *(far left).* A long run of pipe between joints should be supported every 4 feet. Screw the base of the support clamp to the ceiling, attach the clamp to the pipe with the bolt provided and adjust the tension of the strap with the nut. A similar clamp *(left)* is used to support a vertical run of cast-iron pipe. It is attached the same way and should be spaced at 12-inch intervals. A third kind of clamp, attached at floor level on vertical runs, is shown on page 84.

Extending Existing Pipes for New Fixtures

An extra lavatory. A wet bar for a recreation room. A mud room for a gardener. A darkroom sink. A washing machine in a handy location. Such conveniences are easy to provide by applying the techniques of cutting and connecting pipes, described on the preceding pages, to extend an existing plumbing system to a spot where you would like to install a fixture. Only a few basic steps are involved: locate the existing supply lines and the "stack" that serves as drain and vent; connect new piping to them; run the piping to the desired point and put on the new fixture.

For these kinds of extensions, plastic or flexible copper pipes are the simplest to use, mainly because they help eliminate the complications—and hazards—of soldering joints in confined spaces. The choice of materials may be restricted by local codes; it is not restricted by the kind of existing piping to which connections must be made—adapters are available to join any type of piping to any other type. To simplify the job further, you can generally run new piping outside the wall, where it is easy to work on, then conceal it with cabinets or shelves *(page 89)*.

Even the most judicious choice of materials for the new piping is unlikely to eliminate all soldering. If you must use a torch in a confined space, place asbestos sheeting behind the piping to shield surfaces from the flame. Many plumbers wet down the area before they start.

There is one general limitation on the addition of fixtures to an existing system: The new drainpipe must enter the existing drain stack at a point low enough to make waste flow downhill, but not so low that the new fixture's trap will be sucked dry. These requirements translate into maximum lengths for a new drainpipe *(page 85)*, whether it is simply added to an existing drain *(below)* or given its own connection to the stack *(opposite, top and bottom)*. In most communities, a separate vent pipe is not required for a fixture added this way; check your local code to be sure.

In the simplest type of addition—a sink installed to share a trap and drain with another—the connections are made with slip-joint piping; the method is explained below. The sharing of a trap places limitations on such an installation: no more than three sinks can be interconnected, and they must be close to one another.

The two other types of installation sketched require a modification of the drain stack, generally located inside a "wet wall," which is thicker than most house walls to accommodate the large stack pipes. The techniques involved in both extensions are similar, and the instructions on the following pages apply to both, though some illustrations show only the scheme at bottom right.

Three Schemes for Adding a Sink

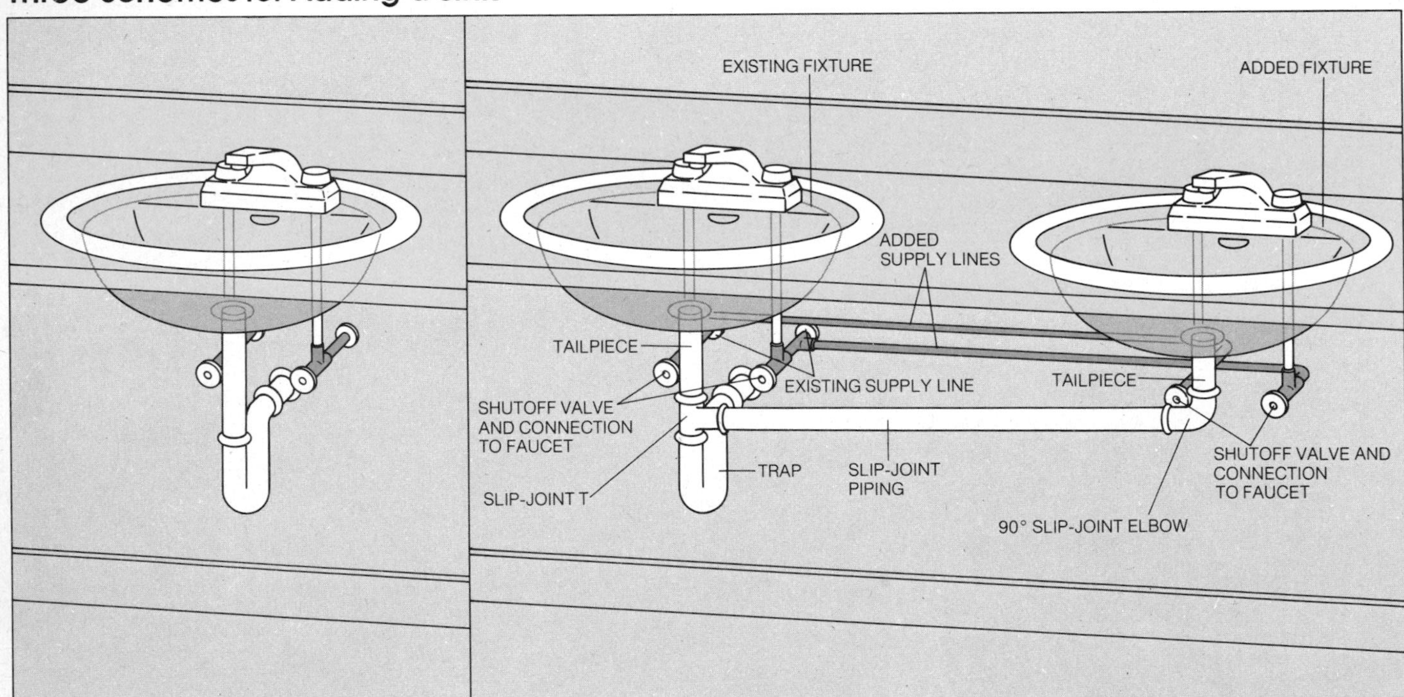

Side-by-side fixtures. The easiest way to extend plumbing requires no work on the large drain stack—the new sink is simply tied into the trap of an existing fixture (shown above before the addition was made) so that both empty into the stack together. Locate the new fixture as described on page 85, but with its drain hole no more than 30 inches away from the existing fixture's drain hole and no more than 6 inches higher. Remove the tailpiece of the existing fixture and install a slip-joint T above its trap. Connect this T to the drain holes of both fixtures with slip-joint piping—a tailpiece for the existing sink, a tailpiece and a 90° slip elbow to the new sink. Remove the existing fixture's shutoff valves, install Ts behind them and replace the existing shutoffs. Then run piping from the Ts to the shutoffs and tubing for the new sink *(pages 92-95)*.

Back-to-back fixtures. If an existing sink or lavatory drains directly into a drain-vent stack in the wall behind it, as illustrated in the drawing below left, a new fixture can be installed back-to-back with the old one without running pipes along the wall. Drain and supply connections are made at the fixture location by replacing fittings already there, as described on the following pages. The Y-shaped section of the stack to which the lavatory is connected is replaced with a "sanitary cross, tapped." (Sanitary means there are no internal ledges to trap waste; cross means the fitting has two inlets, one for each fixture drain; tapped means these inlets have internal threads for fixture drains.) Supply connections are made by installing new T fittings on the existing supply lines, and running pipes from these Ts to shutoff valves for the new fixture.

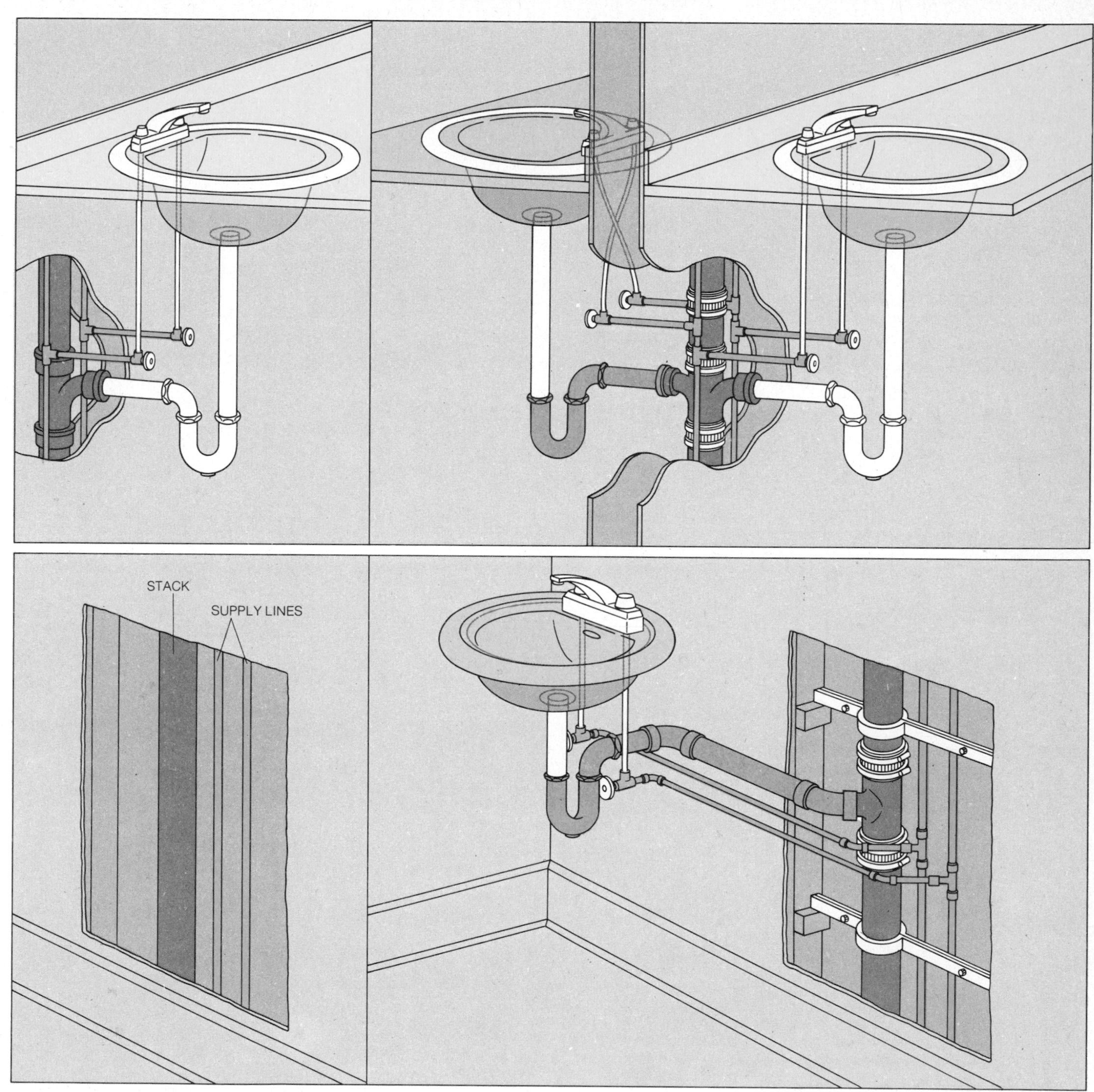

A fixture anywhere. This method enables you to install a fixture at any point close enough to the stack to permit proper drainage, regardless of the location of existing fixtures. You must cut out a section of the stack, install a sanitary TY with tapped inlet there and connect it with drain piping to the fixture trap. Similarly, Ts are installed in the supply risers and connected by new piping to the fixture shutoff valves. Both drain and supply lines can be brought outside the wall near their connections to the stack and risers, then run along the wall to the new fixture.

Locating the Old Lines

The first step in remodeling a plumbing system is to locate the existing pipes hidden in the wall. The vertical drain-vent stack, which determines the location of added fixtures, is not difficult to find. It will project as a vent above the roof, and it will have an exposed cleanout at its base. It is the biggest pipe in the house—at least 4 inches in diameter—and in older houses, it will probably be cast iron. It generally runs almost straight up and down, with only minor jogs.

The hot and cold supply lines may be alongside the drain stack, but in many houses they twist around in the wall in unexpected ways. The best way to track them is to turn on the water, one faucet at a time, and listen for the flow with your ear against the wall. If necessary, drill small exploratory holes—carefully, to avoid penetrating a pipe—until you can hear the water. If you locate one, the other is likely to be within 6 inches of it. Once you have found the supply lines, turn off the water at the main shutoff valve and drain both of the lines by opening the lowest faucets before you begin opening the wall.

Do not be afraid to open a large hole in the wall. A rectangle that spans the space between two studs—usually about 16 inches—will make the work easier, and will be no more difficult to close than a small opening if you make the patch with a matching rectangle of plasterboard.

You must make sure that a drain stack is solidly anchored before you cut into it. Slippage of even an inch will be enough to break the roof seal at the top of the vent and cause a leak. If the drain stack is cast iron, it will also be very heavy. To anchor the stack, install additional stack clamps *(bottom right)*. The one above the section to be removed will support the weight of the stack above; the one below will immobilize the lower stack while you cut. Leave both in place when the job is finished.

1 Gaining access. When you have verified the stack location by drilling an exploratory hole, use a keyhole saw to cut a hole large enough so that you can (a) see where you will tie into the stack—the point where an existing drain enters is sometimes chosen—and (b) locate the adjacent wood studs in the wall. Then insert a steel tape ruler to find how far the stud edges are from the hole. Mark the locations of the stud edges on the wall.

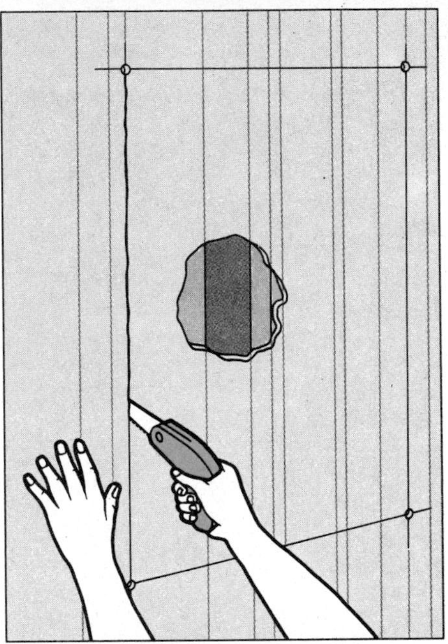

2 Opening the wall. Drill starter holes 24 inches apart vertically just inside each stud line, then cut the rectangle thus marked—24 inches high from stud to stud—with a keyhole saw. Cut only to the stud edges; cleats will be nailed later to the studs to form a nailing lip for a wall patch. If this opening does not uncover the supply lines, locate them and make a similar hole over them.

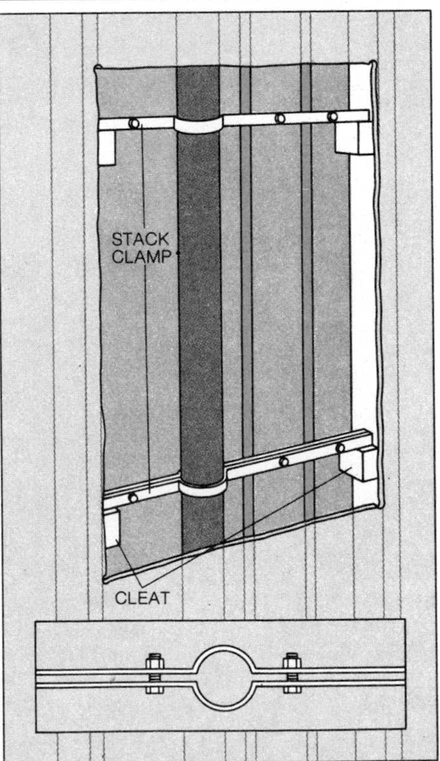

3 Anchoring the stack. To support heavy cast-iron drainpipe while you work on it, place a stack clamp—two shaped pieces of strap steel held together at the ends with bolts *(inset)*—just above and below the section you will tie into. (Similar clamps, installed in much the same way, are available for copper and plastic drain stacks.) Put one strap in back of the stack and the other in front, then tighten the bolts. Support the ends with wood cleats nailed to the sides of the studs, positioning each cleat flush with the front edge of the studs. If you can reach the space beneath the roof, check to see if there is an exposed hub in the stack there. If there is, you can wedge and toenail 2-by-4s under it to give additional support. Be careful not to jar the stack as you toenail the 2-by-4s; you may break the flashing seal that guards against leaks.

A Plan for Piping

You have to juggle a number of factors in establishing the exact location of a new fixture, and a plan marked directly on wall and floor is essential. The major consideration is the angling of the drainpipes for proper removal of waste.

A drainpipe smaller than 3 inches in diameter (such sizes are generally used) must slope downward ¼ inch for each foot of length. However, at the lower end of the pipe—the connection to the drain stack—no part may be below any part of the U bend in the fixture trap; if it were, it would act as a siphon, emptying the trap and rendering it ineffective. These two requirements combine to set limits on the lengths of drainpipes—the so-called critical dimension *(table, right)*. No similar stipulations apply to supply pipes, but they should slope the same way, for convenience in draining them.

At the fixture end of the piping, the relative positioning of faucets and drain is the consideration. When you buy the fixture, be sure you get the "fixture cuts," or "roughing-in dimensions," generally a template that indicates where to locate pipe ends for connections to the new fixture. The supply and trap marks can be shifted on the wall, so long as you stay within the critical distance and pitch the pipe at the correct angle. The drain-hole mark indicates how close you can come to the wall; you can set the fixture farther out if you like.

How you use the fixture marks to plot connections to stack and risers depends to a certain extent on the kind of installation. The illustration below indicates the process when you add a TY at any point on the stack. This method gives you maximum latitude. Mark the fixture cuts first, then extend the trap mark along the wall to the stack, sloping downward ¼ inch per foot: that indicates where the center of the new stack TY should be.

To install a sanitary cross for back-to-back fixtures, start your layout from the center of the cross, sloping drainpiping up ¼ inch per foot to the new fixture location if it is not directly behind the old one. This gives you the point at which the trap must empty into the drain. If you mount a new fixture side by side with an existing one, you must still pitch the drainpipe at ¼ inch per foot, and the new fixture's drain hole must be within 6 inches vertically of the existing one.

The Critical Distances

Drainpipe diameter	Maximum distance to stack
1½ inches	4½ feet
2 inches	5 feet
3 inches	6 feet

Marking the Connecting Points

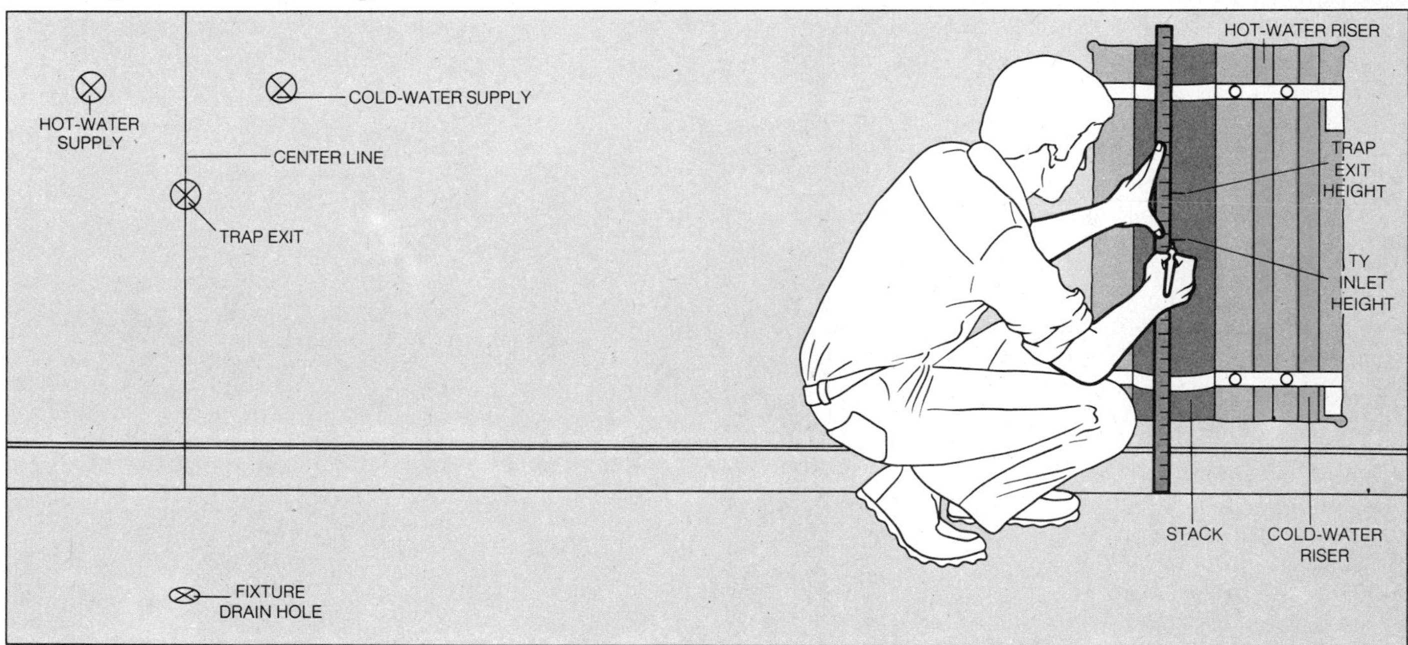

Locating the fixture. If you are using a TY to connect a new fixture's drainpipe to a stack, draw a vertical line—called a center line—on the wall where you want the new fixture to go, but no farther from the stack than the critical distance, as given in the chart at top right. Using the center line as a guide, indicate the positions of the trap and the hot- and cold-water connections on the wall. These marks set the points at which fixture connections—stop valves and trap exit—are to meet the supply lines and drainpipe. Mark the drain-hole location on the floor.

Mark the height of the trap exit on the stack itself and measure the horizontal distance from the center line to this mark. From the height of the stack mark, subtract ¼ inch for each foot of the horizontal distance, and mark the lower height on the stack to indicate the spot for the TY inlet. Draw a line on the wall between the trap-exit mark on the center line and the TY mark on the stack to indicate the correct position and pitch of the drainpipe. In the same way, use the supply marks to find the points that will be the centers of the new Ts in the supply-line risers.

Reverse the sequence of these steps if you are making a back-to-back installation, and draw a line from the existing stack and riser connections to the location of the new fixture.

Making the Connections

With the positions of the new fixture and its pipes established, you are ready to make the connections. The procedures are mainly determined by the kind of piping you must tap into. Start with the drain stack *(right)*. If you are installing back-to-back fixtures, remove the existing connection and cut out the TY fitting in the stack, replacing it with a sanitary cross. Otherwise simply cut out a stack section and install a TY. Cut a copper or plastic stack with a hacksaw, a cast-iron stack with a special pipe-cutting tool. The replacement fittings are easy to install, since they are clamped, soldered or cemented into position, then fitted with adapters if necessary.

Supply-line connections to copper or plastic risers are made similarly, by installing new Ts. If your house has steel or brass supply pipes, the procedure is more complex, requiring the removal of the entire length of a riser between two fittings *(opposite)*, and its replacement with a section of copper piping. In most cases you will find such fittings close above and below the point you want to tap. However, if the nearest riser fittings are a floor above and a floor below, you may have to open walls in several rooms.

In some cases you can avoid such complications by tapping supply lines at a distance from the fixture—there is no restriction on the length of supply connections, as there is on drain connections. For instance, you can take advantage of exposed connections in the basement, where you can use tap-on or compression fittings *(page 73)* without concern about a hidden leak. In the plans shown, both drain and supply run from the stack and risers out of the wall and along the wall; this turn calls for two 45° elbows (for a cross) or one (for a T or TY).

The remaining steps are routine connections made in the open, where you have plenty of room to work. Keep hot and cold supply lines separated by at least 6 inches. And support all piping with appropriate hangers—those for copper should fit snugly to minimize vibration noise, while those for plastic should be a bit loose to accommodate expansion and contraction.

Getting into the Stack

Cutting out a section. To install a TY in a cast-iron stack, set the TY against the stack with its inlet at the level you have marked for it, and mark off the levels of the top and bottom of the fitting. Cut the stack at these two points, using the pipe-cutting tool as explained on page 80.

If the stack is copper or plastic, make the lower cut about 8 inches below the inlet center, using a hacksaw. If you are installing a cross for back-to-back fixtures, just disconnect the drainpipe from the existing TY. Cut the stack about 3 inches above and below the existing TY.

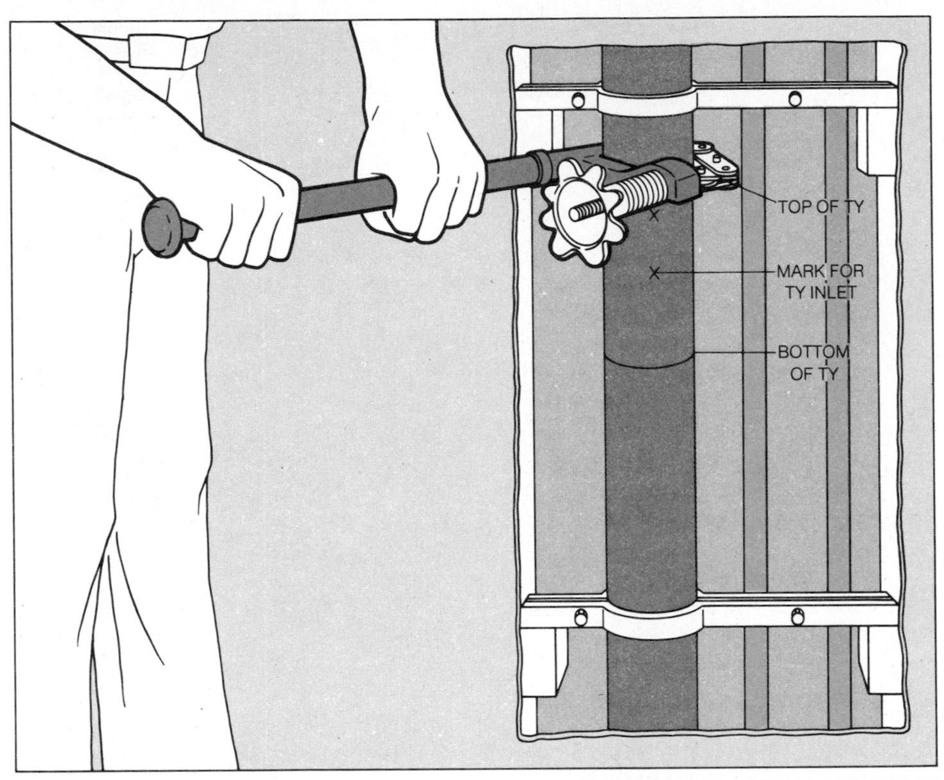

A connection in cast iron. To install a hubless TY with a tapped inlet, slip a rubber sleeve and a clamp onto each open end of the stack and set the TY in place; then slip the sleeves and clamps over the joints, turn the TY to a 45° angle with the wall and tighten the clamps *(right)*. Assemble a hubless cross between two lengths of hubless pipe used as spacers, so that the spacers and the cross together fill the section of stack you have removed and the new cross is at the level of the old TY; install the cross with its inlets at right angles to the wall. Screw the existing drainpipe into the inlet of the cross.

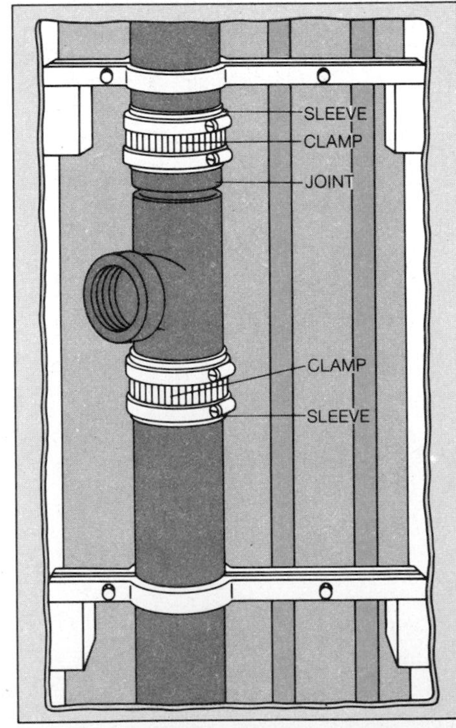

Tapping Supply Risers

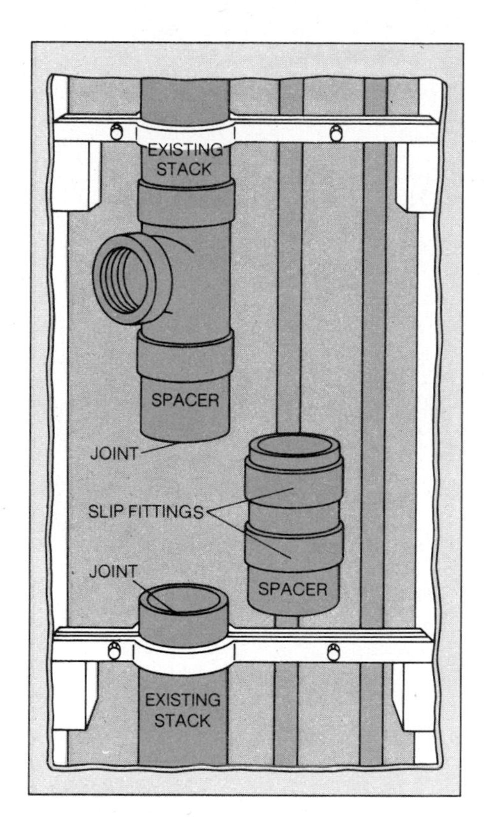

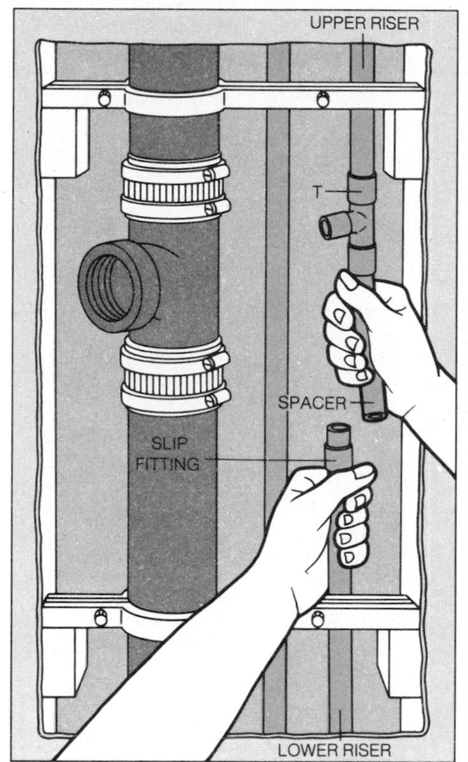

A connection to copper or plastic. Sweat-solder to the top end of the cut stack; angle the fittings as you would in an iron one *(opposite, bottom)*. Solder or cement a short spacer into the bottom of the fitting. Cut a second spacer long enough to fill the remaining opening and slide two slip fittings over it *(above)*. Set this spacer in place, slide the slip fittings over the joints, and solder or cement them.

Connections to copper or plastic. Shut off the supply. If the riser is flexible tubing, cut out a section just long enough so the cut ends fit a T fitting or a cross, then bend the tubing carefully to get the fitting on. If the riser is rigid piping, cut out an 8-inch section and sweat (copper) or cement (plastic) a T onto the upper riser. Slide a slip fitting over the lower riser. Cut a spacer to fill the remaining opening, gently pull the lower riser about an inch to the side *(above)* and sweat or cement the spacer to the T. Holding the spacer and the lower riser in line, slide the slip fitting onto the joint and sweat or cement it in place. For back-to-back fixtures, install new Ts on the risers above or below the existing Ts.

If you are tapping the end of a riser, replace the elbow there with a T fitting. A copper elbow can be unsoldered for replacement. A rigid plastic elbow must be hacksawed off.

Connections to steel or brass. Since you cannot unscrew a single length of threaded pipe, cut the riser at a convenient point with a hacksaw or pipe cutter after shutting off the supply. Remove both pieces of the entire length of pipe between two fittings—you might have to take out long pieces, opening the wall in several places—using a pair of wrenches to unscrew the joint at each end.

Install adapters for copper or plastic piping in each steel coupling. Attach a length of copper or plastic pipe to the upper adapter and solder or cement the new T onto it. Cut a length of pipe about 3 inches shorter than the distance from the bottom of the T to the lower adapter, and install this pipe in the adapter. Then cut a spacer to fit the remaining opening and install it, using a slip fitting for the connection to the pipe. Set the T at a 45° angle to the wall. For back-to-back fixtures, install the new Ts above or below the existing Ts; set the new Ts at right angles to the wall.

Extending Pipes to a Fixture

1 Getting outside the wall. Fit the opening of a TY drain fitting and supply-line T fittings with piping long enough to extend just beyond the wall, and slip 45° elbows onto the ends of these pipes. (In the illustration, a threaded adapter is included at the TY to connect plastic drainpipe to a cast-iron stack.) Running pipe from a cross fitting that has been installed at 90° to the wall requires either one long-sweep 90° elbow or a pair of 45° elbows—one at the cross inlet and a second just beyond the wall surface.

2 Running the drainpipe along the wall. Following the guideline you have marked on the wall, loosely assemble drainpiping to reach the fixture location. Attach a 90° elbow at the end, and complete the assembly with a spacer and a trap; the end of the trap must lie directly over the drain mark on the floor. Prop the entire assembly in position with bricks or scraps of lumber.

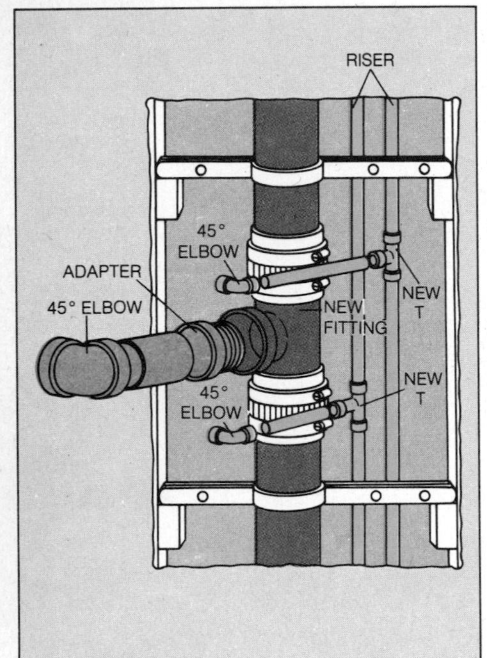

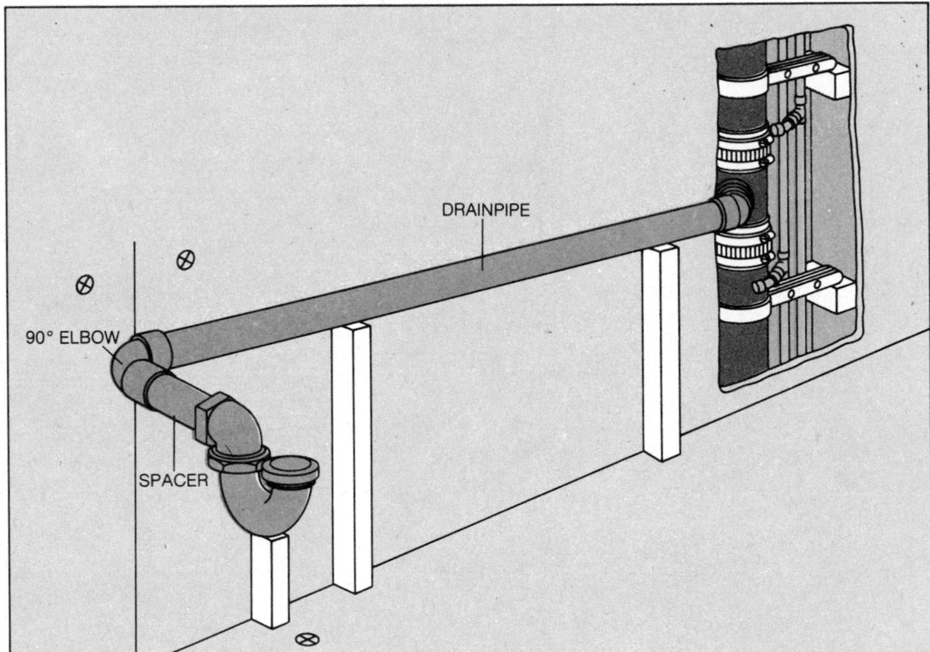

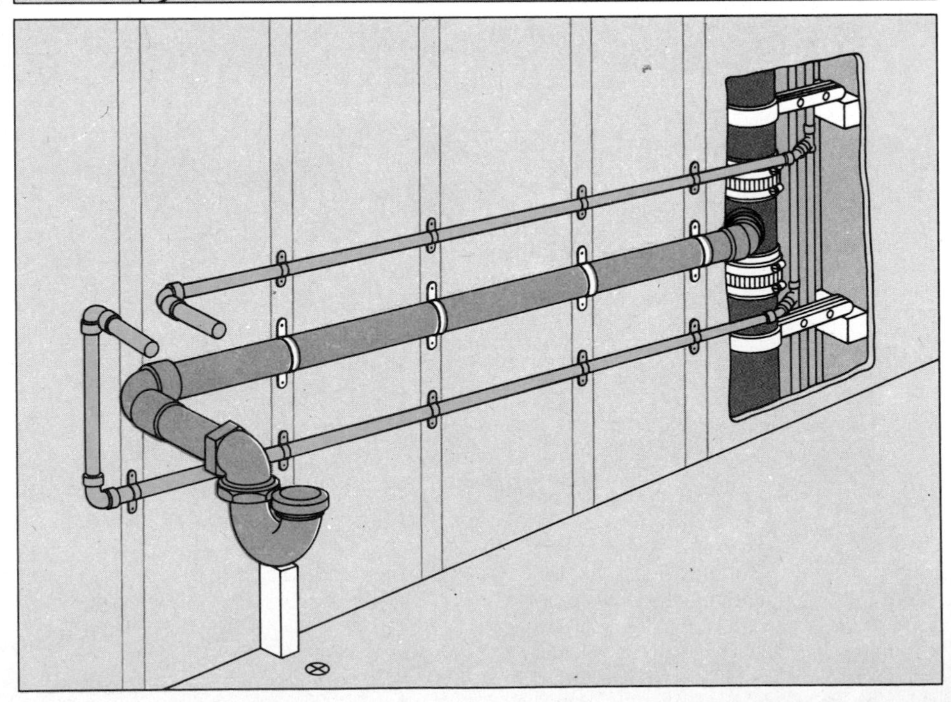

3 Anchoring the pipes. When you are sure everything fits in the drain assembly, sweat or cement the joints. Then anchor the pipe to every stud along its length with metal straps.

Run and anchor supply pipes as you did the drain; their pitch, however, is not crucial. Fit the ends with 90° elbows and short spacers.

The fixture can now be installed and connected —to the trap with slip-joint piping from its drain hole, to the supply through shutoff valves and tubing to the faucets *(pages 92-95)*. Test the system before patching the wall.

Finishing Off around Pipes

1 Adding nailing strips. To provide a lip for attaching plasterboard, nail 1-by-2 strips to the sides of the studs. Make the front edge of each strip flush with the front of the stud.

2 Patching the opening. Cut a piece of plasterboard to fill the opening neatly and cut three strips for the drain and supply pipes. Slip the patch between the pipes and the wall, then push it along the wall and into place over the opening; nail the plasterboard to the strips, dimpling the nailheads. Cover the joints with a thin layer of joint cement, and fill the openings beyond the elbows and the holes around the elbows with thick coatings of additional cement. Cover the outer edge of the plasterboard patch with joint tape and a second layer of cement, then sand these joints smooth. At this point the exposed pipes can be concealed *(below)*.

Concealing Exposed Pipes

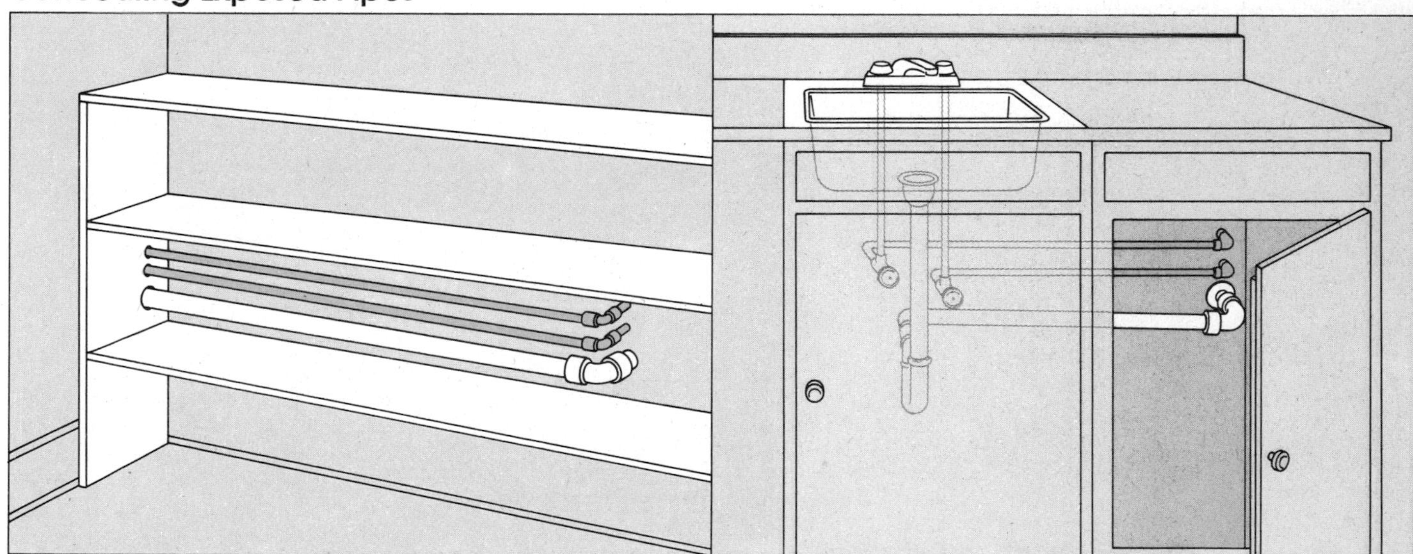

Camouflaging with cabinetry. Although some people paint exposed pipes bright colors and consider them a decorative asset, they are easy to hide, if you prefer, with shelves, boxes or cabinets. Corner piping can be covered with plasterboard nailed to a 1-by-2 frame attached to walls and ceiling; cut the ends of the top frame strips and the long edges of the side strips at a 45° angle to fit the corner. If shelves, resting on cleats, are set in front of the panel, the concealment is even less noticeable. A box, similarly framed of plywood nailed to 1-by-2s, hides a horizontal stretch of drainpipe; bookshelves do the job even more simply—the books on the shelves will hide the pipes *(left)*. Most commonly, the added fixture is simply set into a cabinet *(right)*.

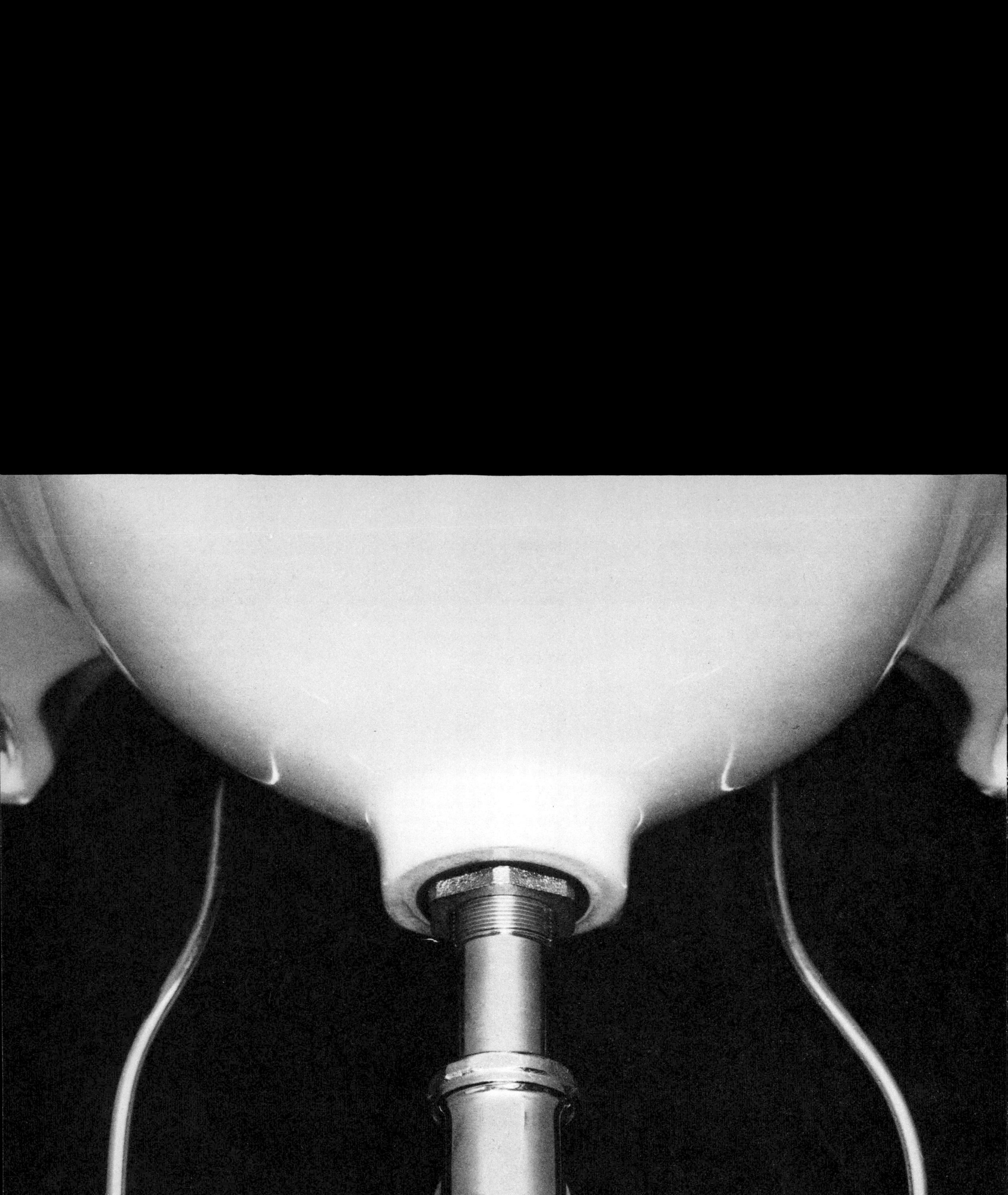

4 Improvements, Indoors and Out

Updating old systems. Flexible chromed-copper connector tubes that bend into whatever arcs you need can take the place of more complex—and costly—arrangements of rigid pipes and elbows to simplify revamping your plumbing. Here the tubes update an old lavatory with new shutoff valves so both the hot- and cold-water lines may be turned on or off at the fixture for future repairs or replacements of the faucet or basin.

If you use piping already in the house to give yourself a head start, you can bring both indoor and outdoor plumbing up to date with a minimum of elbow grease and expense. Small jobs may prove as rewarding as big ones. Simply putting a pressure-reducing valve on a squealing water line *(page 102)* will quiet the whole house—and protect all its fixtures from harmful vibration. Substituting a single-lever faucet for two old-fashioned ones *(pages 94-95)* will make it possible for you to mix water to your choice of temperature without setting down the pan or shampoo bottle you already have in one hand.

Once you get started, seemingly complicated projects often turn out remarkably straightforward. When you replace an old fixture with a new one in the same location, the plumbing connections are simple mechanical fittings; no soldering, threading or cementing is involved. Replacing a rackety water guzzler with a stylish, scientifically engineered toilet, for example, is mostly a matter of loosening nuts with a wrench to take the old tank off the bowl and the old bowl off the floor *(pages 98-99)* and then reversing the process to bolt the new fixture in place. Chances are the trickiest part will be lifting the components—vitreous china is fragile as well as heavy, so you may need a helper to maneuver the bowl and tank safely.

Lavatories call for even less effort to replace. Ready-made fittings go into place quickly. Wall-mounted lavatories are hung like pictures from specially designed brackets packaged with the basins. Modern vanity-type lavatories are simply set into counters or fastened there with screws or bolts *(pages 96-97)*. Neither kind of lavatory is difficult to add to a place where you do not have, but want, one: a darkroom, a bedroom, a mud room; you must extend the drain and supply lines, but if you observe the so-called critical distance *(page 85)* between the fixture and existing pipes, you need not install a special vent pipe.

Outdoor plumbing, which most builders neglect because of the digging involved, may pay the richest dividends of all for the amateur. With a bit of trenching and rudimentary piping, you can treat children to an outdoor shower close to the back door *(page 116)* or indulge yourself in a step-saving lawn hydrant *(page 114)* next to the flower garden. If you want to shoot the works, you can even build in a sprinkling system *(pages 118-123)* and connect it to an electric timer that will turn the whole apparatus on and off automatically.

In all modernizing—additions as well as replacements—keep in mind that you never need feel limited by the materials of your present piping. Adapter fittings *(page 71)* make it possible to join new pipes to old ones. And, in more and more communities, local codes now permit you to use plastic to shortcut your plumbing projects.

Separate Shutoff Valves for Every Fixture

In many older homes, the water supply to the entire house must be turned off every time a plumbing problem arises because there are no individual shutoff valves connected to each fixture. There is no need to put up with this nuisance, since adding shutoffs, or stop valves, is simple if the supply pipes are exposed. Replacing a faulty shutoff is even easier.

You will need either an angled or straight stop, depending on whether the supply pipe—called a stub-out—to which you will attach the cutoff comes from the wall or the floor. The stub-out is generally a short, chrome-plated piece of threaded pipe, but it may be simply an extension of the house plumbing. To make sure you get valves that fit the existing piping, buy them after you have completed the disassembly steps shown here. Take the piece removed from the stub-out to your supplier and have him match it to the valve.

Do not try to reuse the old pieces of piping. Instead, join the stop to the fixture with a flexible connector, ⅜-inch diameter chrome-plated copper tube often known by one brand name as a Speedee. These connectors come in 1-, 2- or 3-foot lengths, which can be cut to the exact size needed. There are three types of flexible connectors, each with a different kind of head, depending on whether it is to be used with kitchen sinks, lavatories or toilets. The drawings show installation of a shutoff on a threaded wall stub-out for a lavatory, but the same procedure applies to floor stub-outs, kitchen sinks and toilets.

Shutoff connections. The angled stop valve is attached to the supply pipe stub-out and to a flexible chrome-plated copper connector leading to the fixture. A straight stop would be used for a vertical connector. The connector shown is a bayonet type for lavatories; flat-headed ones are used for toilets and threaded ones for sinks.

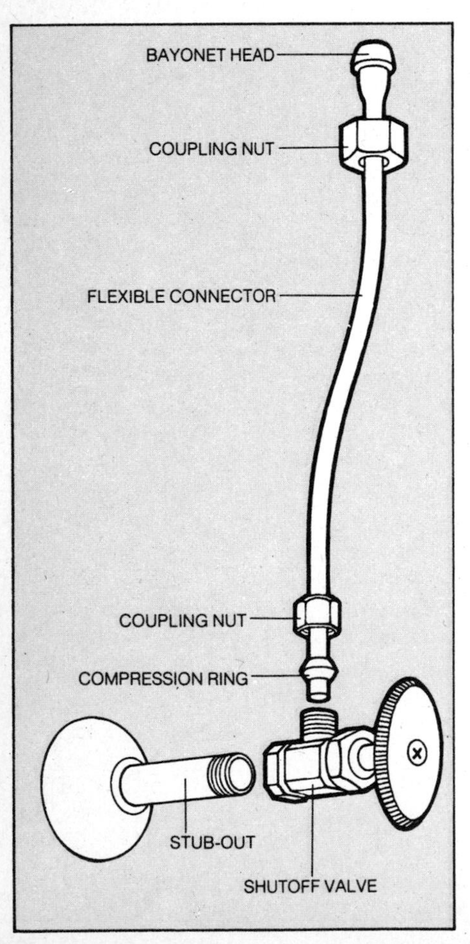

Installing the Valve

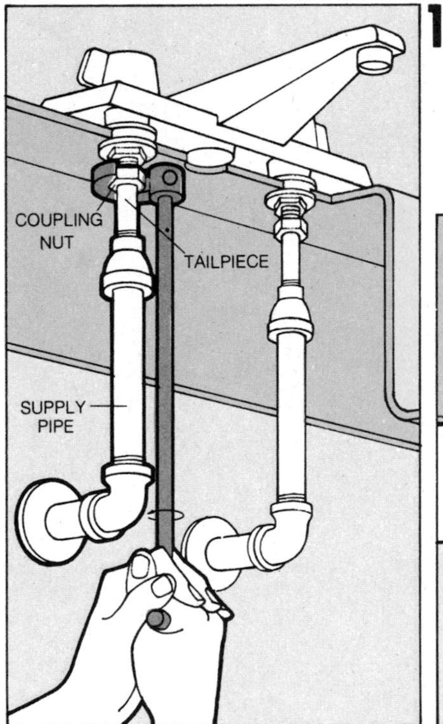

1 Unscrewing the coupling nut. After turning off both hot and cold supplies—at the main shutoff, if necessary—loosen the coupling nut from the fixture with a basin wrench, turning the wrench handle counterclockwise. The nut, once freed, will slide down the narrow, upper section of pipe, called the tailpiece.

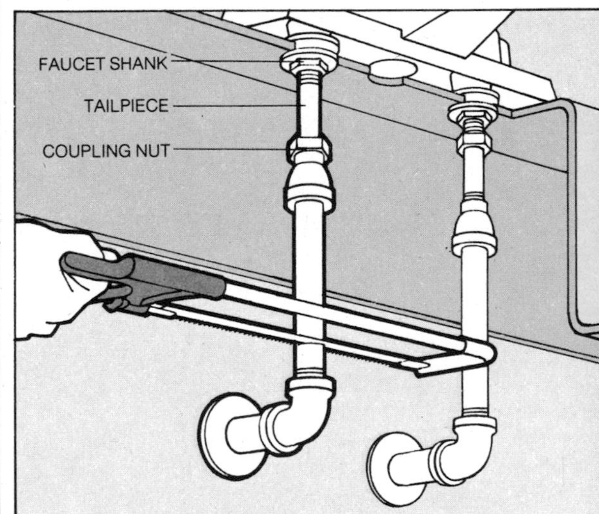

2 Freeing the tailpiece. The surest way to remove the tailpiece is to cut a ½-inch section out of the supply pipe near the elbow with a hacksaw, so that the tailpiece will simply drop out of the faucet shank. It may be possible in some cases to free the beveled end of the tailpiece by pulling the entire pipe assembly down a bit and then gently moving it left or right. But if the pipes refuse to give or there is not enough room to saw, free the tailpiece by loosening the faucet *(pages 94-95)* or toilet-tank ball cock *(pages 56-57)*.

3 Removing the elbow. With galvanized piping, use two pipe wrenches to grasp both the stub-out and the elbow—with its remaining bit of sawed pipe—close to the point where they are screwed together. The wrenches' jaw openings must face in the direction of turning or they will slip. Unscrew the connection, applying pressure on the wrench handles in the directions of the arrows so the back wrench holds the stub-out stationary while the front wrench loosens the elbow. Protect chrome-plated surfaces with adhesive tape.

If the piping is copper, melt soldered joints with a torch or unfasten mechanical connections and remove everything attached to the stub-out. Plastic pipe will have to be cut.

If you have simply slipped the tailpiece out of the faucet shank, you will have to unscrew it from the elbow, using the same two-wrench technique described here, or cut it off near the elbow before you attempt to remove the elbow.

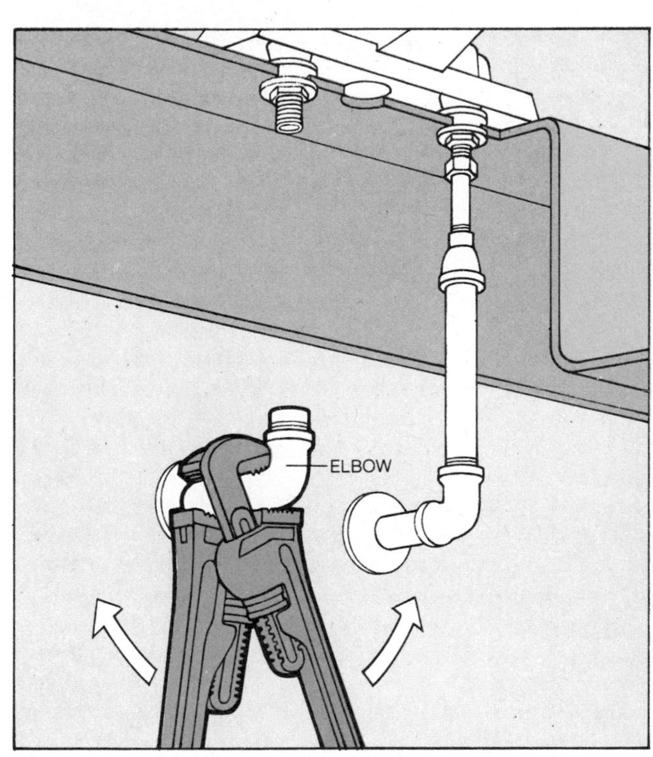

4 Attaching the shutoff valve. Clean the exposed end of the stub-out as for making any joint. The type of fittings you need will depend on the kind of stub-out. For rigid copper pipe use an adapter that can be secured by soldering. Mechanical adapters can be used for flexible copper tubing and some plastics. Rigid plastic pipe requires a plastic adapter screwed to the stop and cemented to the stub-out. On threaded stub-outs, the stop valve is screwed to adapter or pipe after wrapping the threads with a sealing tape. Tighten the connection, using a pipe wrench to hold the stub-out and a smooth-jawed adjustable wrench to turn the stop. The outlet of the valve should be directly below the fixture inlet.

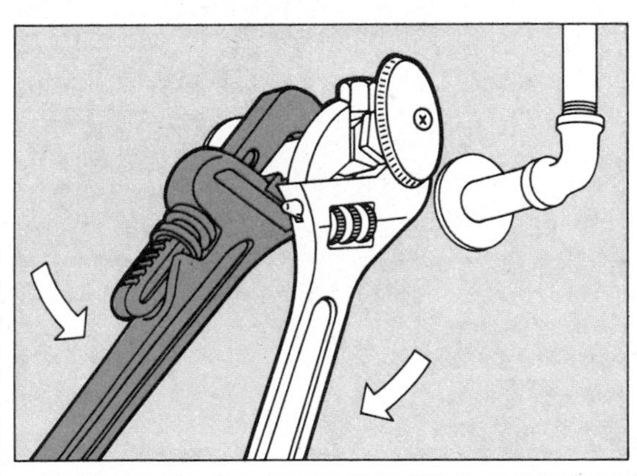

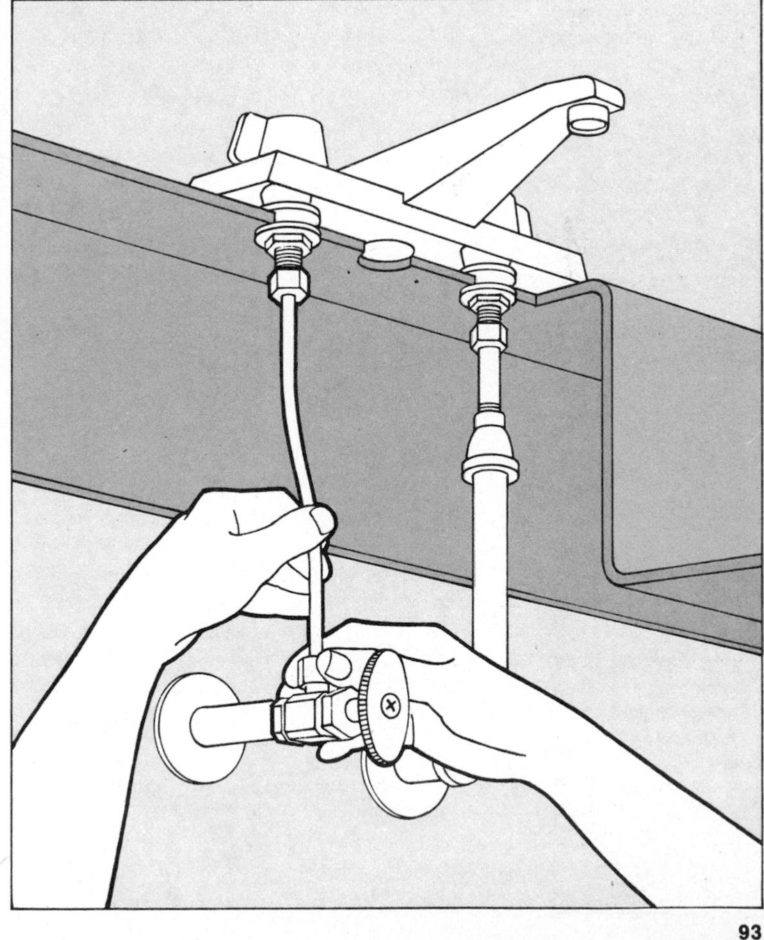

5 Connecting the stop to the fixture. Cut and bend a flexible connector *(right)* to make its head fit the fixture inlet and its end fit inside the stop-valve outlet. Slip the coupling nut over the connector and join the head to the fixture inlet. Screw the nut loosely onto the inlet. Slide the compression nut and ring over the connector's plain end, insert it into the outlet and secure by hand. Finish tightening the coupling nut with a basin wrench, then the compression fitting with an adjustable wrench. Turn the water on, remove the faucet aerator (if there is one) and run the water briefly to clear debris from the pipes.

Updating Old-fashioned Faucets

Installing a new faucet is only slightly more trouble than trying to fix a leak in an old one—and it is well worth the effort. You not only get rid of chronic drips, but also modernize the appearance of your sink or lavatory, especially if you select a single-lever mixing unit. Spray hoses, like the one attached to the unit shown opposite, are a convenience that require no additional plumbing.

To obtain a unit that fits your fixture, measure the distance separating the centers of the holes in the fixture for the tailpiece connections *(below)*. Single-lever faucets are widely available for three-hole lavatories with a 4-inch hole separation and three- and four-hole sinks with 6-inch or 8-inch separation.

If the old fixture has holes that do not match these standard dimensions, use individual faucets as replacements. It is impractical to try to adapt a fixture to a single-lever faucet that is not designed to fit it, since drilling through the porcelainized steel fixture to make new holes is difficult. Filling unneeded holes, however, is simple—chrome escutcheon plates are made to cover them.

1 Removing the old. After turning off both hot and cold shutoff valves, unfasten the tailpiece coupling nuts. You may be able to reach these nuts with locking pliers or an adjustable wrench, but the space between basin and wall is generally so small that a basin wrench is needed. On a lavatory, you next remove the old pop-up drain *(page 48)*. Then unscrew the lock nuts, slide the washers down, pull out the old faucets and remove the tailpieces. Use two pipe wrenches to unscrew threaded joints *(page 78)*, and an adjustable wrench for mechanical connections; soldered joints must be unsweated *(page 75)*.

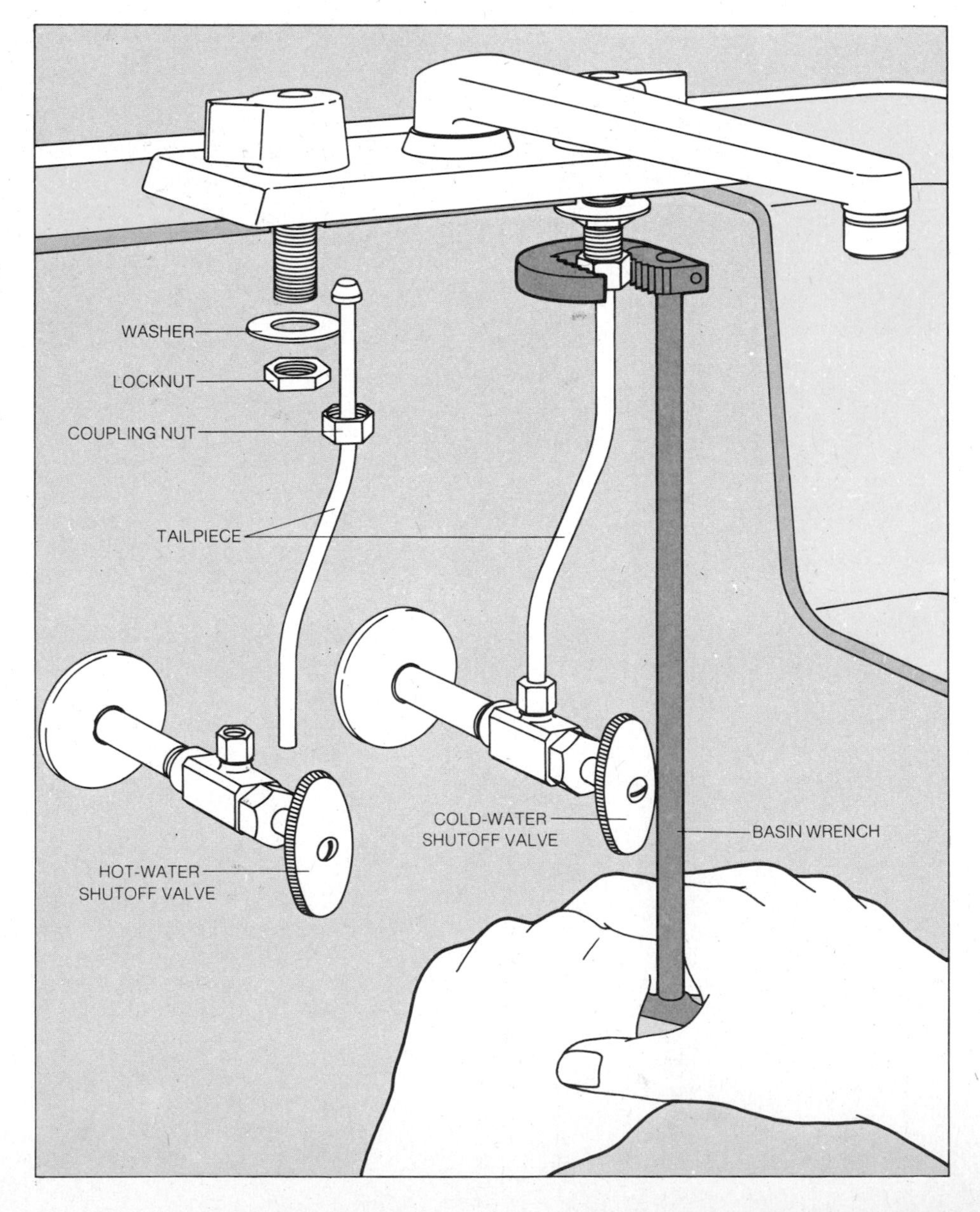

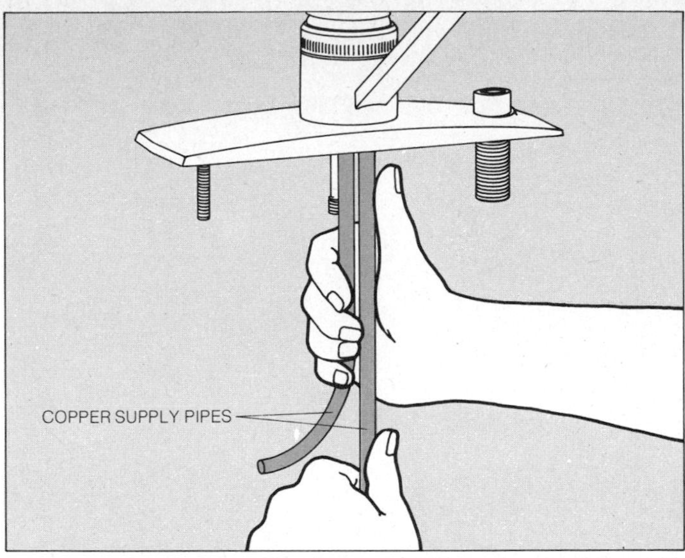

2 Straightening supply tubes. Gently unbend the copper water supply tubes of a new sink faucet so that they will slide easily through the center hole of the sink. Lay the faucet on a flat surface, then straighten each tube with thumb and palm, taking care not to make kinks. On a lavatory faucet, supply tubes are alongside the mounting studs and go through the outer holes; these tubes generally do not need straightening.

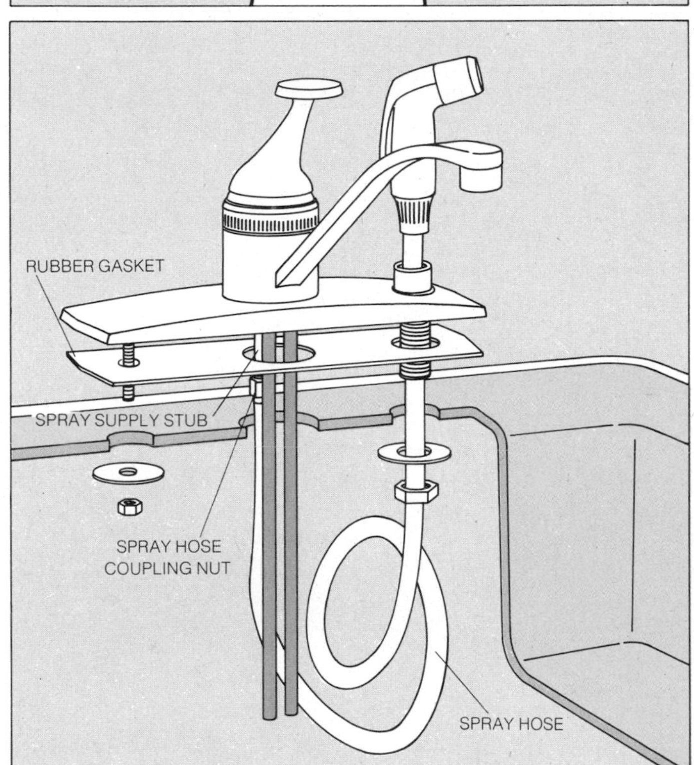

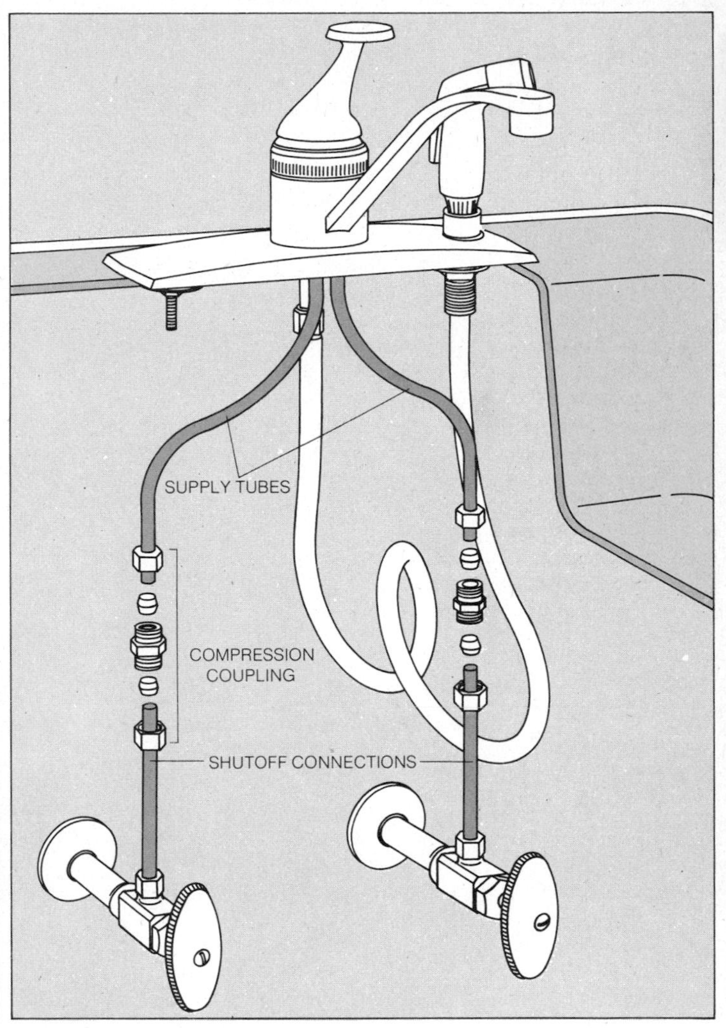

3 Setting the faucet in place. If a sink has a spray hose, attach it first. Slip the hose down through its hole in the faucet assembly, then through its own hole in the sink and finally up through the center hole in the sink. Attach the hose nut to the supply stub on the faucet with a small adjustable wrench. Seal the fixture holes with rubber gaskets or, if your unit does not come with gaskets, plumber's putty.

Slide water supply tubes and threaded mounting studs through the fixture holes. Place washers and nuts on the studs from the underside. Position the faucet carefully and tighten the nuts, first with your fingers and then with a wrench.

4 Reconnecting the supply tubes. Gently bend the supply tubes to line up with the connections of the shutoff valves. (If there are no shutoffs, or if extensions are needed, install them as described on pages 92-93.) Supply tubes and connections are joined with compression couplings. You may need adapters to fit threaded parts to unthreaded tubing. On a lavatory, install the pop-up drain *(page 49)*. When all joints are tight, turn the water back on, remove the aerator from the spout and run hot and cold water hard for several minutes to wash out the lines and check for leaks.

New Styles in Lavatories

Taking out an old lavatory and putting in a modern one is part plumbing and part carpentry, and neither job is difficult. You just disconnect the old faucets and drain, and hook up new ones as shown on pages 92-95 and pages 50-51. The carpentry consists simply of removing the old fixture and mounting the new one.

An outdated fixture may sit on a pedestal that has to be taken out *(right);* others are wall-hung on hangers or angle brackets. Designs vary, but most wall hangers have projecting tabs that fit into slots at the back of the lavatory or under the back rim, or they have pockets that hold special lugs molded into the underside of the lavatory. New wall hangers and angle brackets can be mounted with toggle bolts *(below),* but the wall hangers may then need to be supplemented by adjustable legs—especially if children are likely to climb on the fixture. Angle brackets always require legs for front support. For the most secure legless installation, wall hangers should be mounted on a backing board, mortised into the studs *(opposite, top left).*

A countertop lavatory fits into a hole in the top of the counter; the manufacturer usually supplies a template for cutting the hole. Some bowls, the self-rimmed type, overlap the countertops and are supported by them. A frame-rimmed model is secured with lugs that connect frame, lavatory and countertop *(opposite, bottom).* Unrimmed, recessed lavatories are held up by clips and screws. All must be sealed into the countertop with plumber's putty.

Lavatories can be ordered with holes for either a 4-inch or an 8-inch faucet assembly—which is sold separately and should, in most cases, be attached before the lavatory is set into place. Countertop lavatories usually are installed in cabinets 31 inches high. Wall-hung ones may be mounted with the basin top at any comfortable height up to 38 inches.

Removing the old lavatory. After turning off the water, disconnect the supply pipe. Disconnect any bolts that hold the basin to the top of the pedestal. Lift off the basin. Rock the pedestal back and forth to loosen it from the plaster of paris holding it to the floor. If necessary, wrap the base with a towel to keep chips from flying and, wearing goggles, pound with a hammer until the base comes loose. Scour or sand off the plaster residue remaining on the floor.

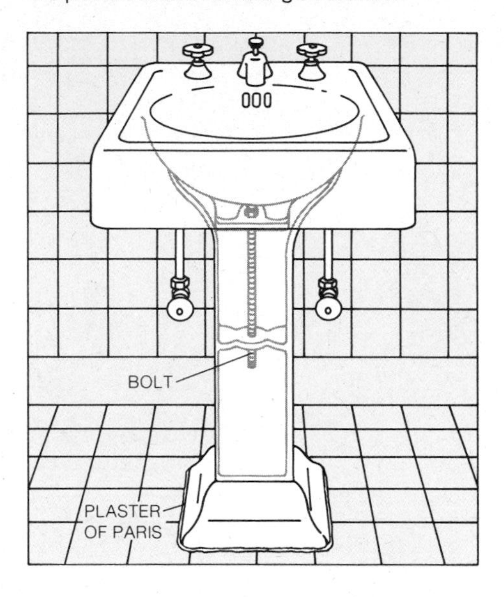

Three Kinds of Hanging Devices

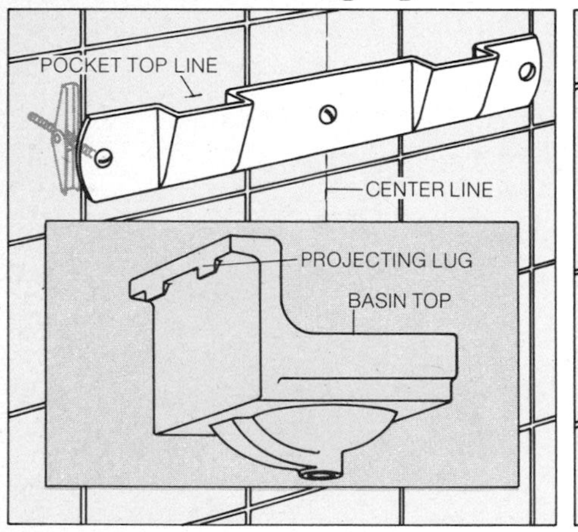

Attaching a pocket-style wall hanger. Draw a vertical.center line up the wall from the middle of the waste pipe. Measure the distance from the top of one of the projecting lugs on the lavatory to the top of the basin *(inset)* and add the desired height from the floor. Draw a horizontal line across the center line at this level. Center the hanger on the vertical line with the tops of the pockets at the horizontal line. Drill through the center hole in the hanger and insert a toggle bolt or screw. Using a level to keep the hanger horizontal, set bolts or screws in the remaining holes. Attach the faucet and drain to the lavatory, then slide the lugs into the hanger pockets.

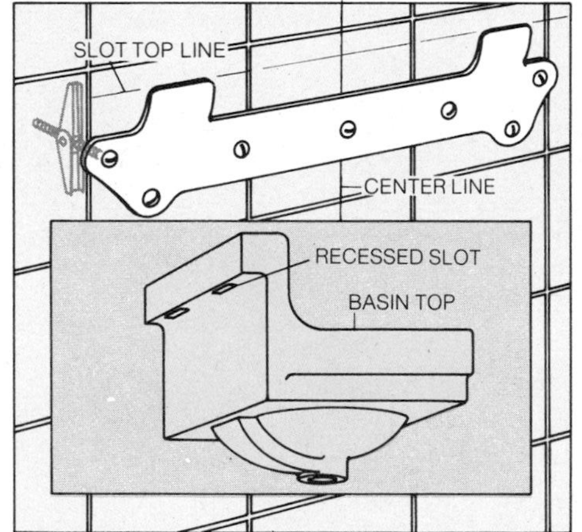

Attaching a tab-style wall hanger. Draw a vertical center line up the wall from the middle of the waste pipe. Measure the lavatory from the top of a recessed slot in the back (or the underside of the top rim, if there are no slots) to the top of the basin *(inset)* and add the desired height from the floor. Draw a horizontal line across the center line at this level. Center the hanger on the vertical line with the tops of the tabs at the horizontal line. Fasten the hanger to the wall and install as with a pocket-style hanger.

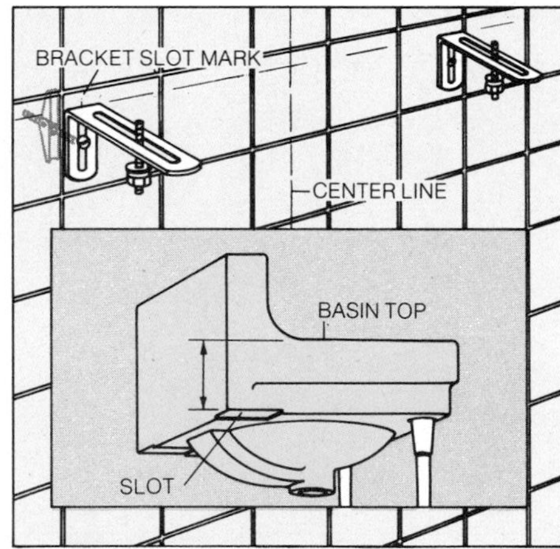

Attaching angle brackets. Draw a center line up the wall from the middle of the waste pipe. Measure from the bottom of the lavatory to the basin top *(inset);* subtract this figure from the desired height from the floor. Draw a horizontal line across the center line at this level. Then measure from the center of the lavatory to one mounting slot and mark this distance from the center on each side of the horizontal line. Place each bracket on a slot mark, aligning the top to the horizontal line. Fasten the brackets to the wall with washers and toggle bolts or screws. Attach the faucet and drain and install the lavatory by inserting bolts and washers through the mounting slots. Add front legs *(opposite, top right).*

Reinforcing a Wall-hung Basin

Installing a backing board. At the approximate location planned for the lavatory, remove enough wall surfacing to expose two studs and form an opening at least 4 inches high. Notch the studs to hold a 1-by-4 board flush with the edges of the studs. Screw the board to the studs, then refinish the wall in front of it.

Position the hanger over the backing board *(opposite, bottom)*, and loosely attach the hanger to the board with one 3-inch screw. Using a carpenter's level to keep the hanger horizontal, insert screws in the remaining holes.

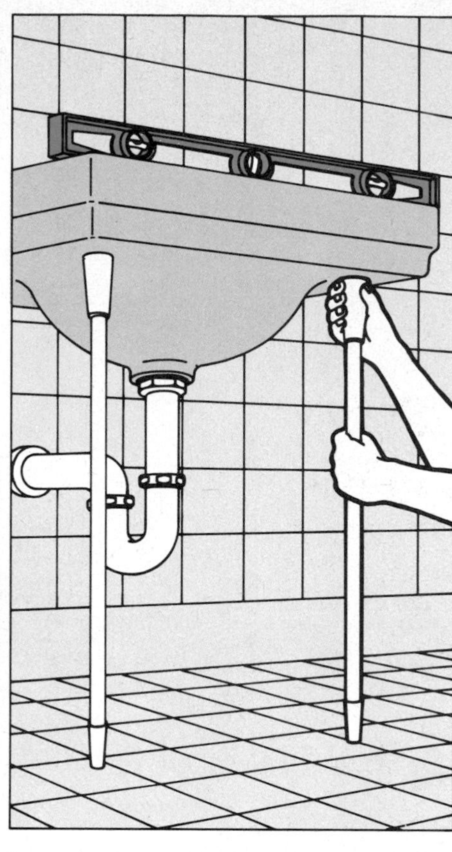

Attaching front legs. After installing the lavatory on its hanger, brace the front with a 2-by-4, if necessary. Insert the tops of the legs into holes at the front corners underneath the lavatory or, if there are no holes, inside the front rim near the corners. Stand the legs upright with the bottom pin tips positioned in the grout spaces between floor tiles if possible. If the pin tip must rest on a tile, use a carbide bit to drill a hole ¼ inch deep to accommodate the tip. Place a carpenter's level on the front lip of the lavatory and carefully screw the adjustable center section of each leg downward until the leg just stands securely and the lavatory is level. Do not overlengthen the legs. Tighten any nuts at the tops of the legs and above the adjustable sections. Remove the 2-by-4.

Installing a Countertop Basin

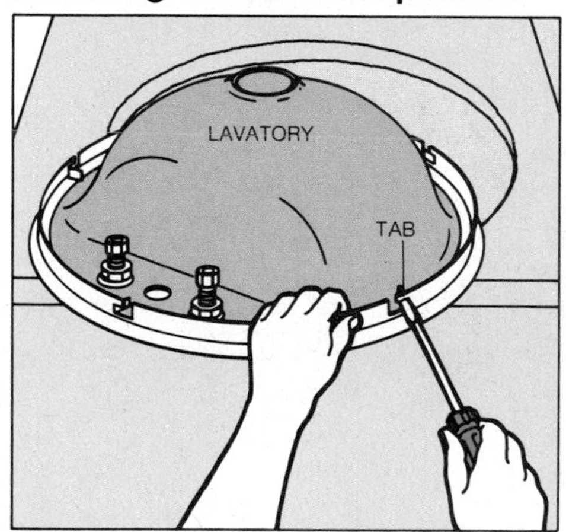

Attaching the frame. Use a template or the bottom edge of the frame to cut a hole in the countertop. Secure the top to the cabinet. Connect the faucet assembly to the lavatory, then turn the bowl upside down. Run a bead of plumber's putty around the inside flange of the frame rim and set the bowl into the frame. Using a screwdriver, punch in the tabs on the frame to hold the lavatory temporarily in place.

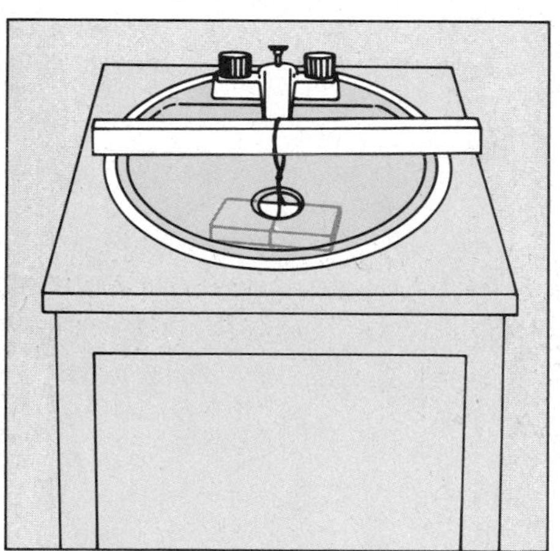

Setting the bowl in place. Apply a ¼-inch-wide strip of plumber's putty around the edge of the hole in the countertop. Set the framed lavatory into the hole, pressing it down until the excess putty squeezes out. To support the lavatory while you attach the mounting lugs, place a 2-by-4 across the cabinet top and run a wire from the board through the drain. Tie a wood block to the bottom of the wire, then twist the wire around the block until it fits snugly against the bowl.

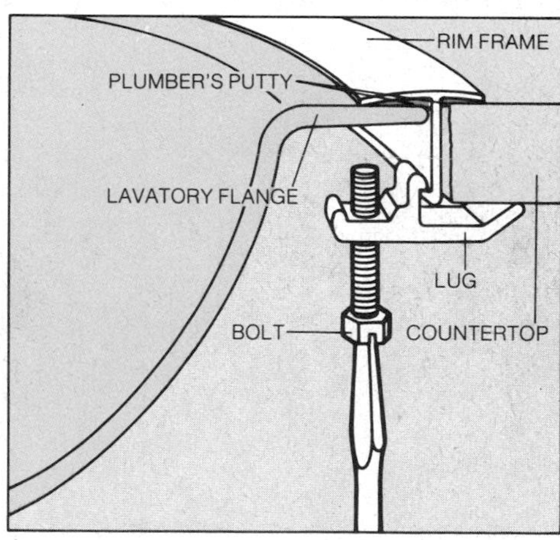

Anchoring the lavatory. Fit lugs over the bottom lip of the frame at evenly spaced intervals and secure the frame to the underside of the lavatory and countertop with bolts. Remove the board and wire turnbuckle. Wipe up the excess putty around the edge of the frame. Install the drain and hook up the water and waste pipes.

Replacing a Toilet

Installing a new toilet from scratch takes a good deal of plumbing expertise, but replacing a damaged or old-fashioned one with an up-to-date model is a job any householder can complete in an afternoon. No major plumbing work is involved. The replacement bowl fits over the existing drainpipe and floor flange. And the tank can be connected to the existing water supply pipe, even if you are replacing an ancient wall-mounted tank *(right)* with a modern toilet that has the tank mounted on the rear of the bowl. Water-saving models, built with lower tanks and narrower bowl traps to release about a third less water at each flush, are installed in exactly the same way as the conventional type.

The only critical factor in buying a new toilet is the so-called "rough-in distance" from the wall behind the toilet to the center of the drainpipe. This is a simple measurement *(Step 1, above, right)* but be sure to make it carefully; your plumbing supplier will need it to provide a new toilet that will fit into the same space as the one you remove.

The internal mechanisms of the new tank will already be installed and you will also get the necessary washers, gaskets and hardware for fitting the tank to the bowl. You may need to buy hold-down bolts, for securing the tank to the floor; ask your supplier. Also, buy a new wax gasket *(Step 4)* for sealing the bowl to the drainpipe, and a small can of bowl-setting compound to make a watertight seal between bowl and floor.

You will probably need to reroute the water supply pipe to connect it to the tank. This is easily done with flexible connecting pipe. But be sure to ask for a toilet-tank supply pipe; its fittings are different from those for a sink or lavatory. If the old toilet did not have a shutoff valve, now is the time to buy and install one *(pages 92-93)*.

The job requires only a few tools: a spud wrench (or, if you do not have one, a large pipe or monkey wrench), screwdriver, carpenter's level, tape measure and putty knife.

1 Finding the rough-in distance. With the old bowl still in place, locate the hold-down bolts that secure the bowl to the floor. Measure from the center of the bolts to the wall behind the bowl. This is the rough-in distance—it determines the exact location of the concealed drainpipe in relation to the wall. (If the bowl has four hold-down bolts, measure from the rear pair.) The rough-in distance of the new bowl may be somewhat shorter than that of the old one, but it must never be longer or there will not be enough space for the new fixture.

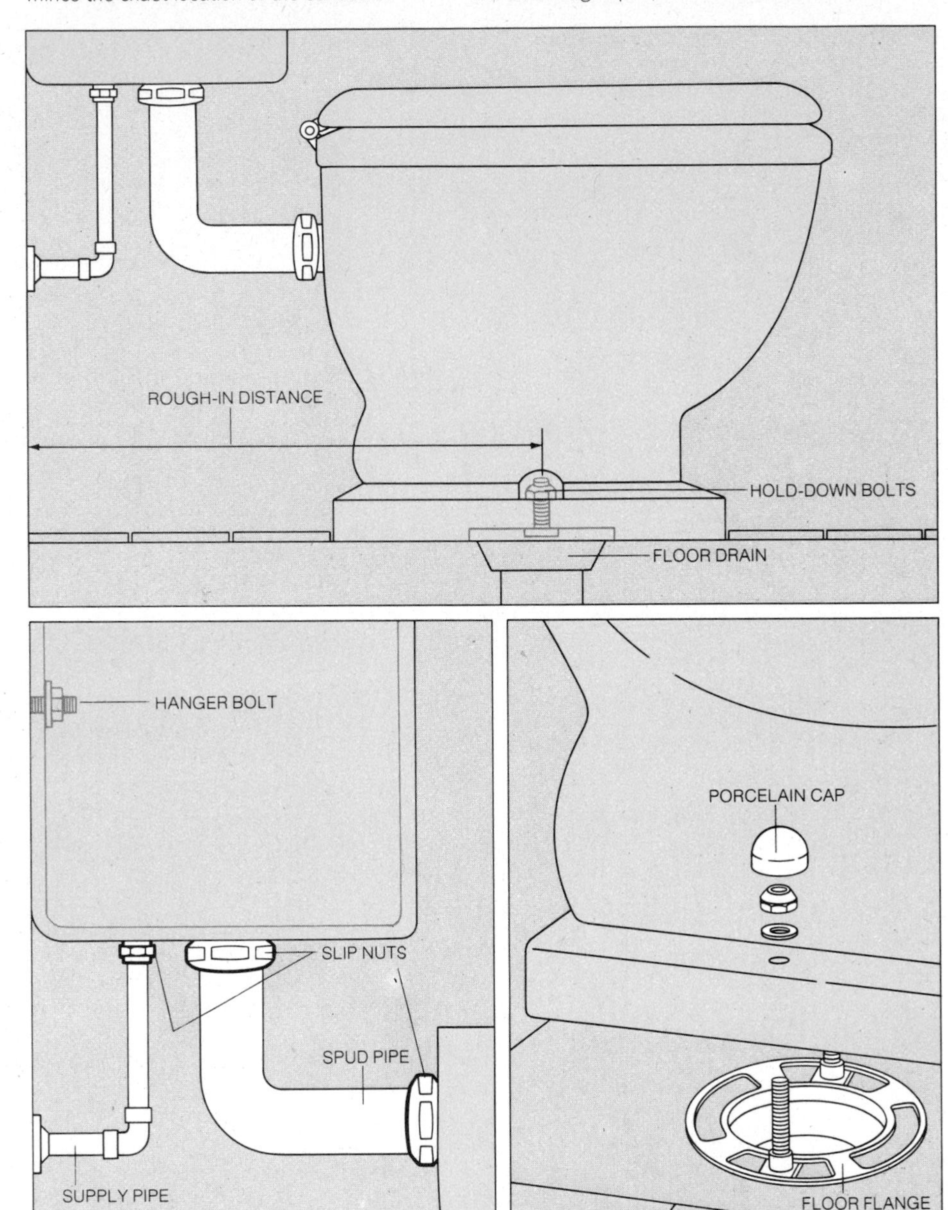

2 Removing the tank. Shut off the water supply, flush the toilet and sponge out the remaining water in the tank and bowl. Disconnect the supply pipe by loosening the slip nut at the tank. If the tank is wall-mounted, use a pipe or spud wrench to loosen the slip nuts on the spud pipe. Remove the spud pipe. While a helper supports the tank, unscrew the nuts from the hanger bolts and remove the tank from the wall. If the tank is mounted on the bowl, remove the nuts from the bolts in the upper rim of the bowl. Then lift the tank off the bowl.

3 Removing the bowl. Unscrew or pry off the porcelain caps of the floor bolts, and remove the hold-down nuts and washers. Badly corroded nuts may have to be soaked with penetrating oil. To break the seal between the bowl and the floor, grasp the bowl and twist or rock it back and forth. Carefully lift the bowl straight up off the bolts and set it aside. Stuff a rag into the drainpipe opening to keep sewer gas inside the pipe and to prevent debris from falling in. Using a putty knife, scrape away the remnants of the old wax gasket or putty from the floor flange.

4 Installing the wax gasket. With the new bowl upside down, place a wax gasket around the water outlet (called the horn). If the floor flange is recessed below floor level, you will need a wax gasket with a plastic sleeve. Install the gasket with the sleeve facing away from the horn.

5 Installing the new bowl. Remove the rag from the drainpipe opening. Turn the bowl upright, position it over the floor flange and press down with a twisting motion to tighten the seal between bowl and drainpipe. Use a level to be sure the bowl is not tilted. If necessary, insert shims, improvised from thin sheet metal, under the base of the bowl to make it level and to keep it from rocking. Screw the washers and hold-down nuts onto the floor bolts, which should emerge through the base of the bowl. Do not tighten the nuts or attach the bolt caps yet.

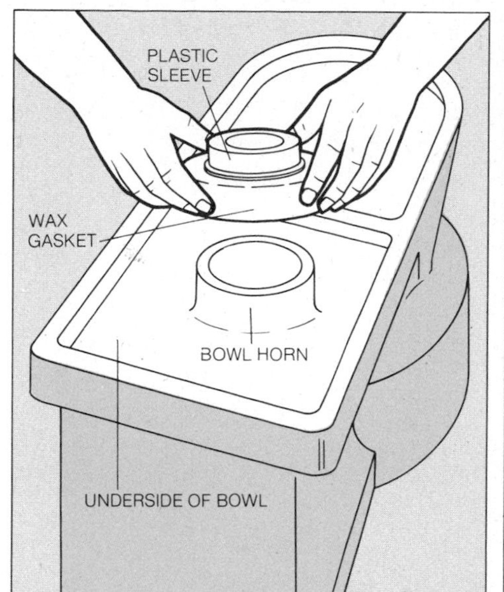

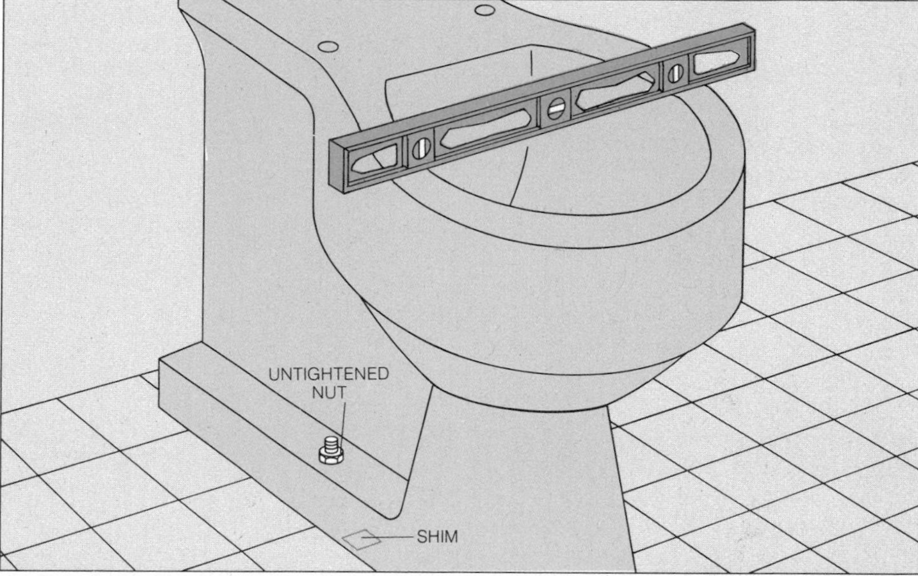

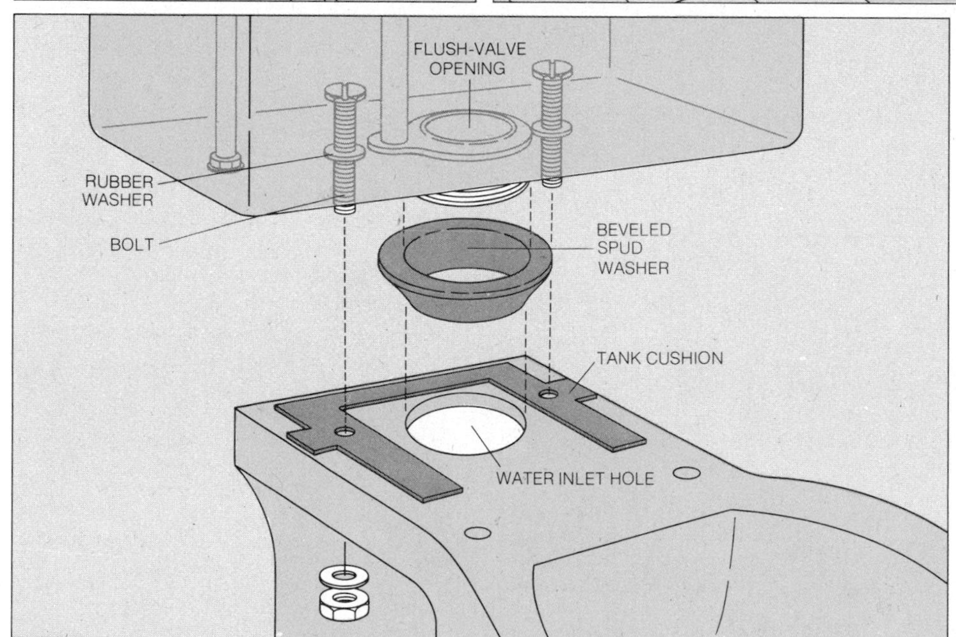

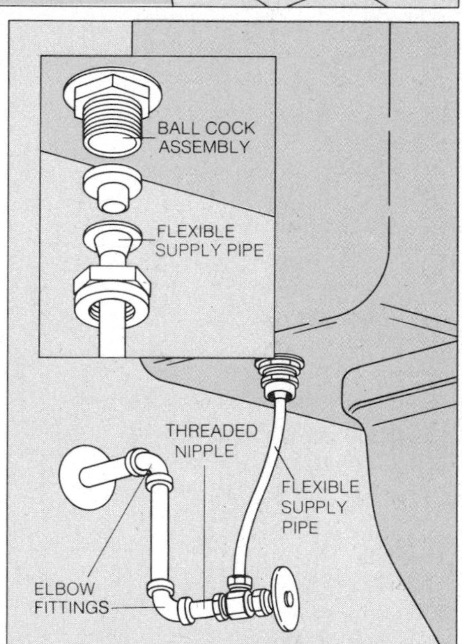

6 Installing a bowl-mounted tank. Fit the beveled rubber spud washer around the flush valve opening at the bottom of the tank. Then place the rubber tank cushion over the rear part of the bowl so that the two holes in the cushion align with the bolt openings on each side of the water-inlet hole. The two bolts (cushioned with rubber washers) that you insert through the underside of the tank will fit through these openings when you position the tank over the bowl. Fasten the bolts with nuts and washers where they emerge underneath the bowl rim. Adjust the alignment of the toilet so that the tank is parallel to the wall (they may be a few inches apart). Check the bowl to make sure it is still level and does not rock. Then tighten the hold-down nuts and bolt caps at the base of the bowl. Seal the base to the floor with toilet-bowl setting compound or plaster of paris, trimming away the excess with a putty knife. Attach the seat and cover *(page 60)*.

7 Connecting the water supply. If your new tank is lower or farther from the wall than the old tank, you must readjust the fittings for the supply pipe or install new fittings. To bring the old fittings to the desired level, screw an elbow onto the wall stub-out (if there is an elbow in place, turning it 90° may suffice), then use a threaded nipple to connect a second elbow. If the old toilet lacked a shutoff valve, install one *(pages 92-93)*. Install the flexible supply pipe, connecting the flared end to the ball-cock shaft *(pages 56-57)*. Tighten the connections and turn on the water.

A Replaceable Water Filter

Tap water that has an unpleasant taste or odor because it contains rust, chlorine, sulfur or organic material can be purified by installing a water filter with a replaceable charcoal or carbon core. A filter is easily inserted in a water supply line—either in the main line near where it enters the house, or in a pipe attached to a major fixture, such as the cold-water faucet in the kitchen that supplies water for cooking and drinking. Where you put it affects the way it is installed. For installation on a horizontal water line *(below)*, you can buy a kit with prefabricated parts; for the other installations *(below, right)*, you will have to assemble your own filter mount from standard pipes and fittings. Whichever arrangement you use, the filter must be installed in an upright position with one or more valves isolating it from the system; the valves are needed so that the water flowing through can be shut off while the filter core is being replaced.

The filter core lasts from six months to a year, depending on the quality of the water and the water flow—one in the main line must be replaced more frequently. When the core does have to be replaced, shut the water off, unscrew the filter body from the cap, pull out the old filter core and put in a new one. Then thread the filter body back onto the cap and turn on the water.

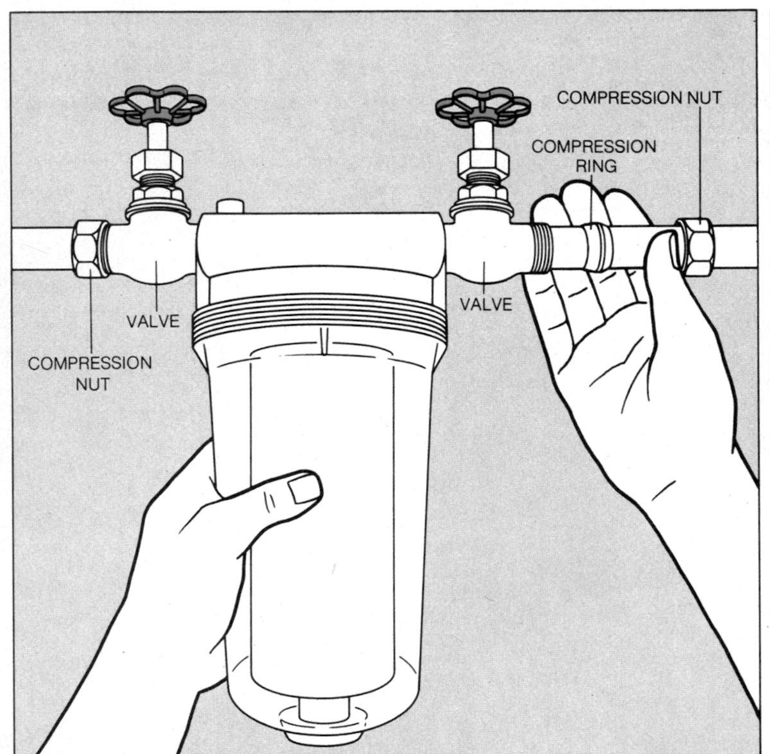

A filter in a horizontal line. To install a water filter on a horizontal water line, first shut off the water at the main valve; then cut a length out of the pipe where the filter is to go, using a template that is provided in the valve kit. Thread a valve onto each side of the filter cap and tighten them until each is in an upright position. Slide a nut and compression ring over each pipe end, fit the filter and valves over the pipe ends, and tighten the nut and compression ring onto each valve as you would for a standard compression fitting *(page 73)*. When replacing a used filter core, shut off both valves.

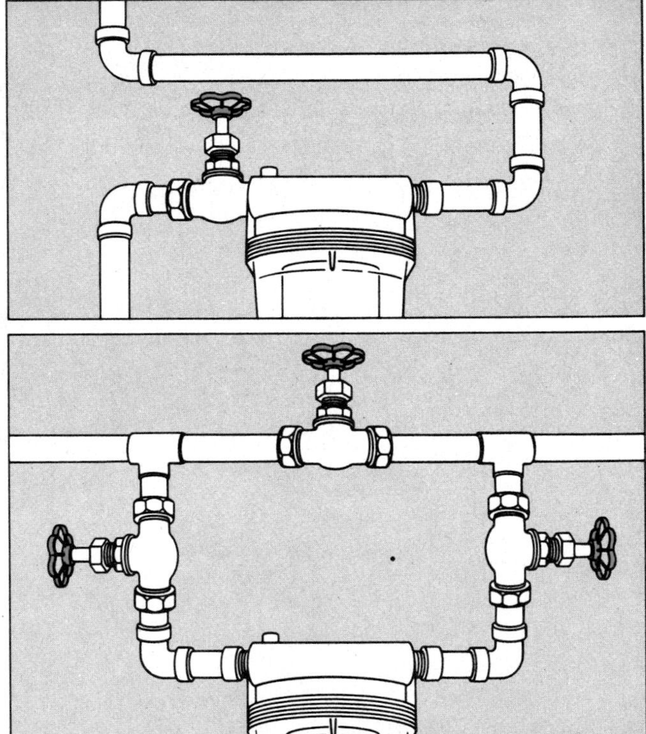

A filter in a vertical line or bypass. In a vertical pipe *(top)*, shut off the water and cut a 4-inch section. Install a loop and insert the filter in its lower leg, placing a valve on the inlet side. For a bypass installation *(above)*, insert a loop as illustrated; then opening the top valve and closing the side valves permits the filter to be replaced without shutting off the water supply.

Hand-held Showers

The hand-held shower—often called a telephone shower because of its resemblance to a telephone receiver—is an easily installed, inexpensive luxury. Some types simply replace a standard shower head. More versatile is the type illustrated, which adds a spray source to the existing head. Its flexible metal hose is attached to a special diverter valve that is part of either the shower *(right)* or the bathtub spout *(below)*. The tub-spout diverter is an integral part of the spout itself, while the shower-head diverter is a separate valve that is meant to be installed between the existing shower head and the shower arm.

When buying a hand shower, look for one that can be easily disassembled. Then you can readily get at parts if they should need replacing, and the occasional job of cleaning out sediment and water scale is also greatly simplified. Plastic spray parts help minimize the accumulation of scale-forming minerals.

Installing tub and shower diverters. To put in a shower-head diverter *(above)*, remove the old shower head *(page 43)* and thread the diverter valve onto the shower arm. Tighten the valve with two wrenches until the diverter button is at the top. Attach the new shower hose to the bottom outlet, then thread the regular shower head onto the main outlet of the diverter in the same way it was installed on the shower arm.

A tub-spout diverter is installed the same way as a standard tub-spout replacement *(page 43)*.

Guards against High Pressure

Abnormally high pressure, either in a supply pipe or inside a water heater, causes problems ranging from the mildly annoying to the downright dangerous. Since even the nuisances can eventually damage faucets and fittings, such problems deserve the prompt solutions that the devices on these pages provide.

Water hammer, which is a banging noise caused by vibrating pipes *(page 16D)*, is a phenomenon that often occurs when a valve—like the automatic one on a clothes washer—abruptly stops the pressurized flow of incoming water. You can eliminate it either by installing a ready-made shock absorber *(far right)* or by assembling an air chamber *(bottom)* from standard pipes and fittings. The shock absorber costs more but it is smaller (only 4½ inches high), easier to put in and maintenance-free.

Cavitation, a squeal caused by bubbles in the water escaping through a small orifice *(page 16C)*, is a problem of excessive pressure, common in homes close to a city water tower or pumping station. By installing a pressure-reducing valve *(opposite, top)* near the point where the water enters the house, you can step line pressures of 80 pounds and more down to a manageable 40 pounds.

Steam pressure sufficient to explode a water-heater tank can develop if it is not fitted with a relief valve; some tanks lack these safety devices, but you can install one by either of the two methods shown opposite, left. Replace an old valve when your periodic check *(page 13)* indicates that it is worn out. If the tank lacks a proper drain line, add one *(opposite, right)* to protect yourself from being scalded when you check the valve.

Anatomy of a shock absorber. This device for eliminating water hammer is a small cylinder containing a rubber bellows surrounded by hydraulic fluid, with inert gas filling the space above the fluid. When the hammer-causing valve closes, the pulsing water in the line expands the bellows and the inert gas cushions the vibrations to eliminate the noise. After the pressure subsides, the weight of the hydraulic fluid around the bellows compresses it back into its original position.

Installing a shock absorber. Insert a T *(page 86)* with a ¾-inch threaded outlet into the water supply line just ahead of the problem valve. Apply joint cement to the threads of a 1½-inch-long ¾-inch steel nipple, then screw it into the T's outlet. Thread the shock absorber onto the nipple.

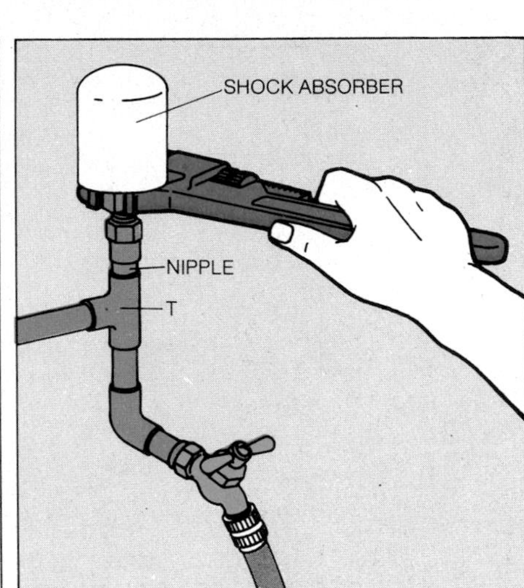

Making an air chamber. The simplest air chamber is merely a capped pipe set into a T near a problem-causing valve. To build a more effective chamber, insert a T in the water line near the troublesome valve and attach a shutoff valve to the T with a short length of pipe. Using another T and more short pipes, as shown, install a drain cock above the shutoff valve and a reducer fitting above the drain cock. The large end of the reducer must be at least twice the diameter of the water line. Attach a pipe at least 2 feet long to the reducer and seal the top with a cap.

When the problem valve on the water line shuts off, incoming water rushes up into the chamber and the air cushions its force. Eventually, an air chamber becomes waterlogged. To restore its effectiveness, empty it by temporarily closing the shutoff valve and opening the drain cock.

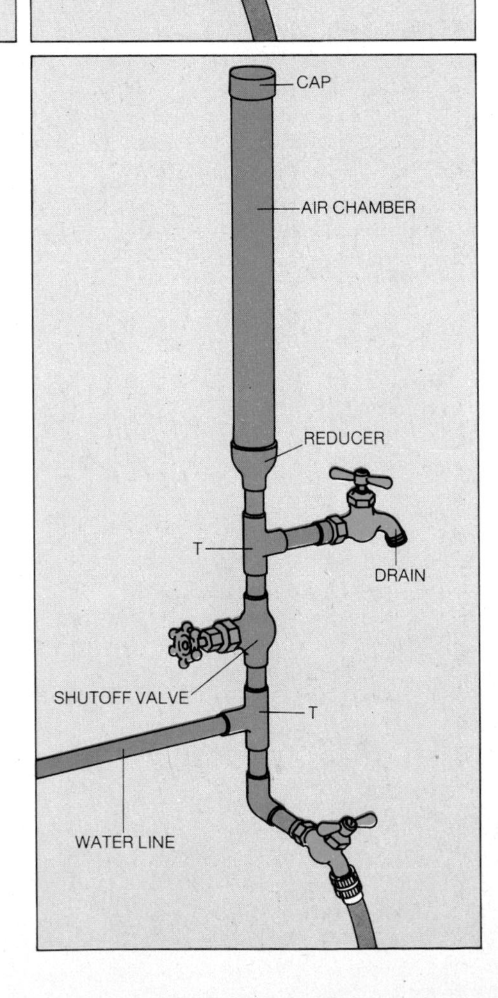

Putting in a pressure-reducing valve. Thread an adapter suited to the material of the water pipe into the inlet and outlet of the pressure-reducing valve. Measure the length of the valve and adapter fittings—including a union *(page 79)* if the line is made of steel. Close the main shutoff valve and cut out a section of the main water-line pipe to the length of the valve and its fittings. Attach the pressure-reducing valve and fittings to the pipe ends and turn the water back on. Turn the adjusting nut at the top of the valve clockwise until the pressure is reduced enough to end cavitation and water hammer, but still supplies adequate flow to the upper floors of the house.

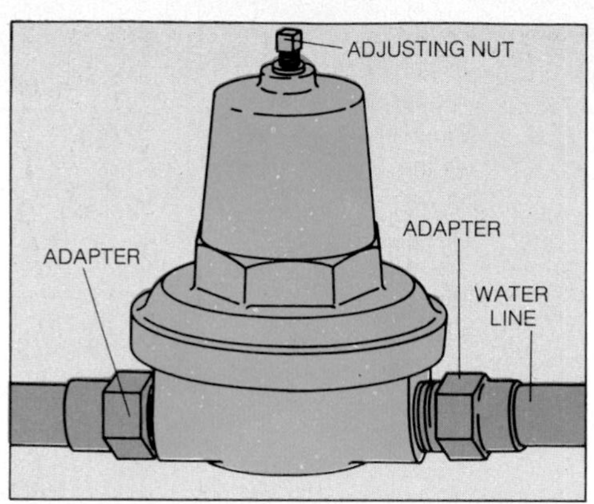

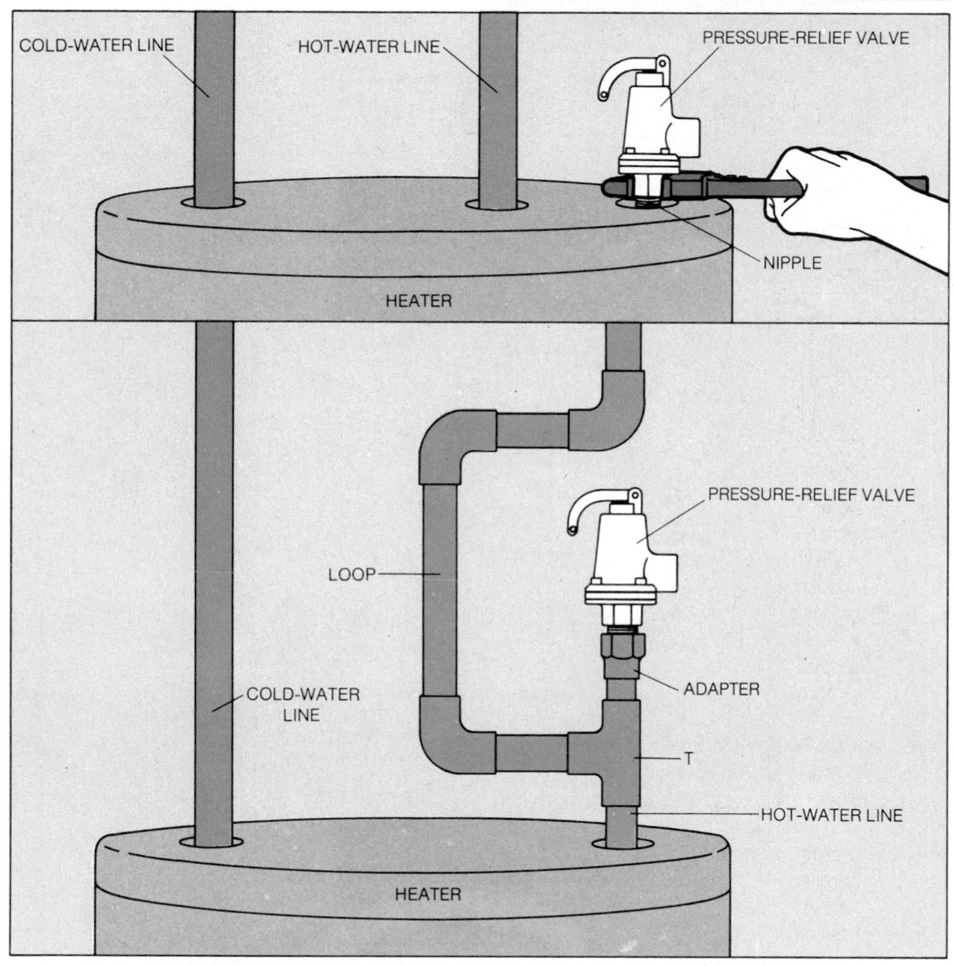

Installing a pressure-relief valve. Shut off the gas or electricity to the heater and let it cool down. Screw a ¾-inch steel nipple into the recessed, threaded ¾-inch fitting on the top of the heater, then screw the pressure-relief valve to the nipple. If there is no recessed fitting, cut a 10- to 12-inch section out of the hot-water line just above the point where it emerges from the top of the heater. Install a T *(page 87)* over the lower part of the line. Attach one end of a threaded adapter to the top of the T and attach the valve to the adapter. Run the hot-water line from the side outlet of the T and reattach it to the line above with a loop *(above)*.

Putting a drain line on a water heater. Order two pieces of prethreaded steel pipe the same diameter as the outlet of the pressure-relief valve—one piece long enough to reach from the valve outlet to just beyond the edge of the heater tank and the other long enough to reach from the top of the tank to within 6 inches of the floor. Thread the shorter pipe to the valve outlet, using two wrenches, as shown on page 78. Attach a 90° elbow to the end of the short pipe, then thread the longer pipe to the bottom of the elbow.

Washers and Waste Disposals

Since clothes washers and dishwashers both need piping to supply clean water, and drain lines to dispose of used water, locate them near a sink or tub. If a clothes-washer drain line is too short to reach a laundry tub, install a standpipe drain like the one below, far right. The dishwasher drain can go either into the sink drain or into an inlet provided on nearly all garbage-disposal units.

Since a dishwasher uses only hot water, it needs only a single supply line; a clothes washer needs both hot and cold lines. Either appliance should have a shutoff valve in the supply line in case the machine needs servicing. Moreover, it is important to shut off the water to a clothes washer whenever it is not in use because the machine itself is connected only by rubber hoses; under constant pressure, such a hose can burst while no one is around to notice and can quickly flood a basement several feet deep.

Two different valve arrangements can be used to control a clothes-washer supply *(below)*. With either, a shock absorber *(page 102)* should be installed in each of the lines to prevent water hammer, which is often caused by the abrupt closing of the solenoid valves in the washer.

Hooking Up a Clothes Washer

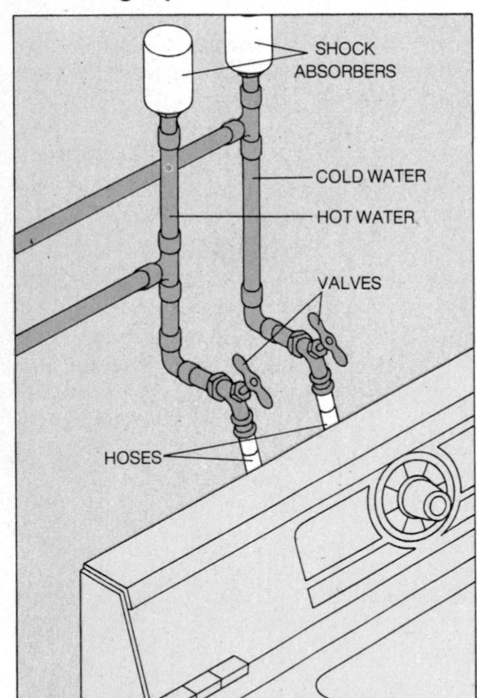

Water lines with separate valves. Locate the hot- and cold-water supply pipes nearest the washer, drain them and install a T fitting in each pipe *(page 87)*. Run a pipe from each T to a point just above the washer. Put another T into each pipe and install shock absorbers *(page 102)* in the side connections of these Ts. Extend the pipes down from the shock absorbers, leaving enough space above the washer for shutoff valves. Install elbows on each pipe and attach shutoff valves having threaded spigots to accept the machine hoses.

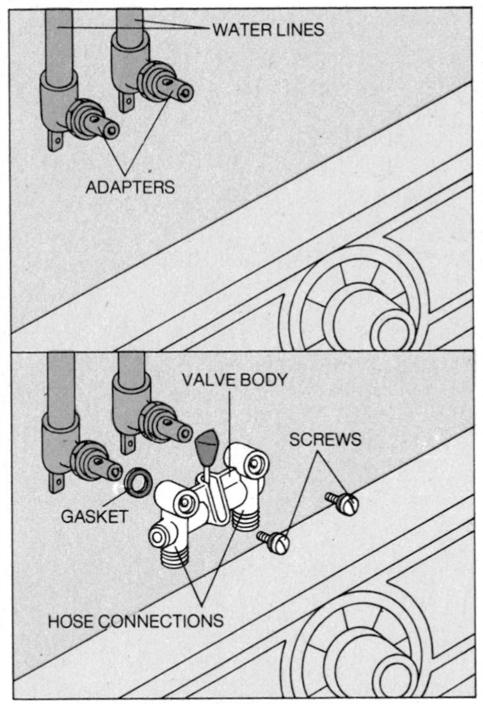

Attaching lever-valve adapters. A lever valve simplifies shutting off the water supply to the washer. Run new hot- and cold-water pipes with shock absorbers, as for the standard valves illustrated at left, to the vicinity of the washer as above. Unscrew the adapters from the valve and sweat one to the end of each pipe, using adapter fittings if your pipes are not copper. Be sure that they are aligned so that the valve end of each valve adapter will slide smoothly into the valve body. Slide gaskets and then the valve body onto the valve adapters. Insert and tighten the attachment screws, then thread the washer hoses.

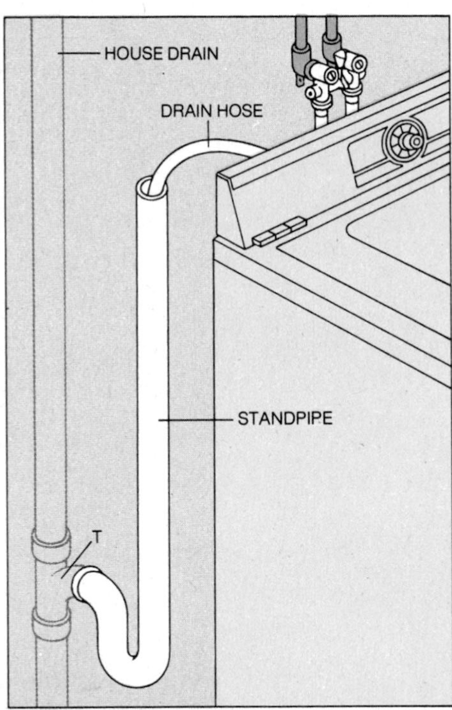

Installing a standpipe. If there is no nearby sink in which to place a clothes washer's drain hose, cut into a drainpipe within reach of the hose and install a T *(page 86)* that will accommodate a standpipe 2 inches in diameter. Standpipes are available, complete with trap, in lengths from 34 to 72 inches. To determine the size you need, check the manufacturer's installation instructions. Attach the standpipe to the T and insert the washer's drain hose into the standpipe far enough so that it will not be pushed out of the pipe by the force of the draining water. Never connect a clothes-washer drain hose directly to a house drain; there must be an air gap to prevent back-siphonage of dirty water into the machine.

Piping for a Dishwasher

Running a water supply pipe. Shut off the hot water and install a T *(page 87)* in the pipe under the sink for copper tubing. Run flexible copper tubing from the T to the water inlet valve on the machine. Insert a shutoff valve in the pipe. Move the machine into position and attach the supply pipe to the water inlet valve of the washer, according to the manufacturer's instructions.

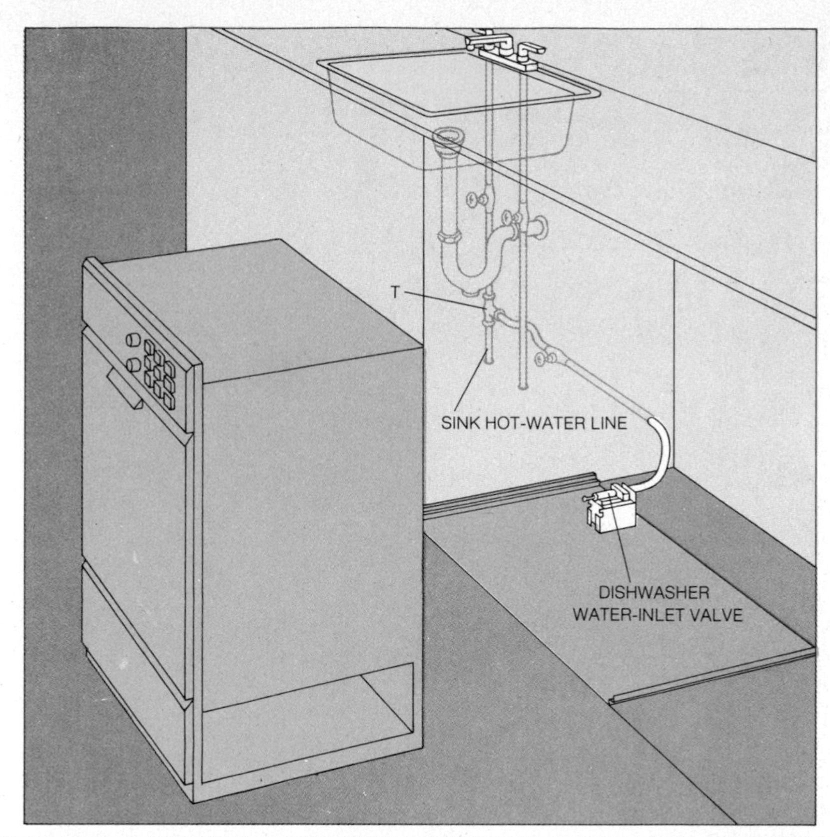

Installing a waste T. A dishwasher drain hose can be permanently connected by inserting a waste T in the sink drainpipe. If your dishwasher drain hose ends with a threaded coupling, purchase a waste T with a threaded side connection. If not, buy a T with a smooth side connection and attach the drain hose to it with a hose clamp.

Remove the tailpiece of the drain, slip the smooth bottom section of the waste T into the trap and secure it by tightening the compression fitting on the trap *(page 73)*. Cut the tailpiece so it will fit into the gap between the top of the waste T and the sink drain. Set the tailpiece into the compression fittings at the top of the waste T and the bottom of the sink drain and tighten both fittings. Attach the hose to the side connection of the T.

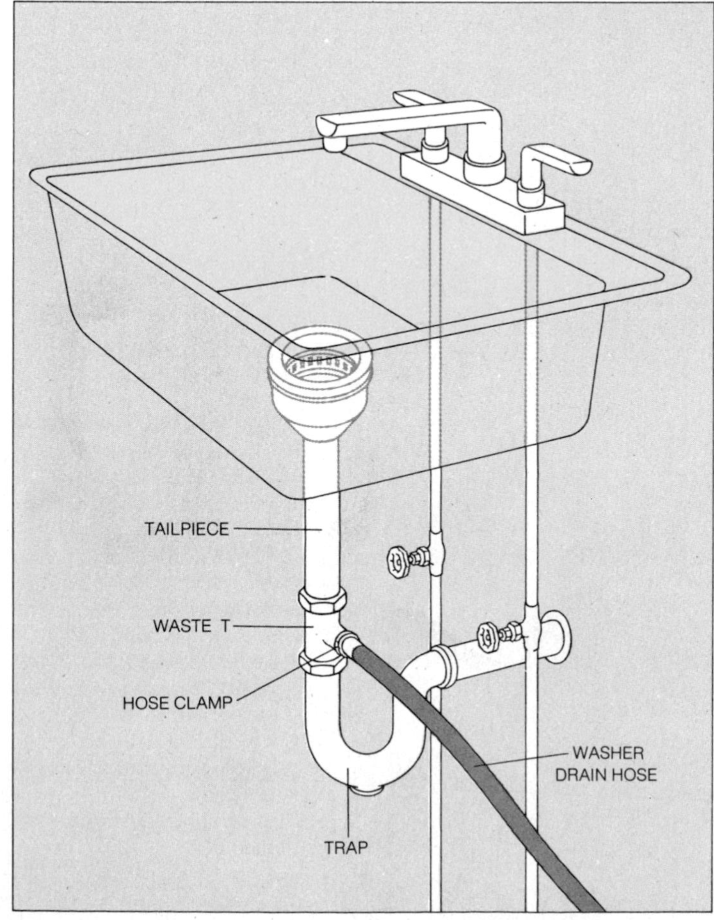

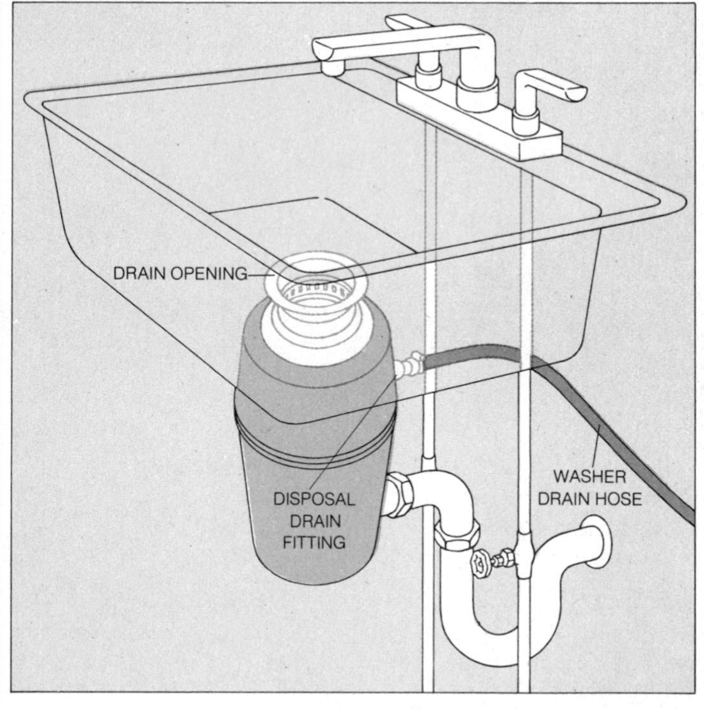

Tying into a garbage disposal. Most garbage disposals have a fitting to which a dishwasher's drain line can be attached. Turn off the electric current to the disposal. Insert a screwdriver through the sink drain into the disposal and pry out the knockout plug that seals the fitting. Slip the end of the dishwasher drain hose over the end of the fitting and tighten it with a hose clamp.

An Add-on Drainage System

New clothes washers, bathrooms, garbage disposers and other water-using appurtenances naturally put an additional burden on the household drainage system—a clothes washer, for example, accounts for 40 per cent of the average family's liquid waste. If the house is served by a community sewer system, the increased flow is of little concern, but it can be a major problem if the drainage goes into a septic tank, as it does in 15 million American homes. There is, however, a simple and relatively inexpensive solution that allows you to put in some new fixtures without taxing—or even tampering with—the septic tank. The solution is to build a seepage pit. This is a separate drainage facility that can dispose of "clean" waste—water that is free of solids or germ-laden material and thus does not require septic-tank treatment.

A seepage pit, or dry well, is nothing more than a lined hole in the ground that collects water and allows it to disperse and be absorbed slowly into the surrounding soil. The hole can be lined with stone, brick, wood or concrete. Digging the hole is a chore. Keep it shallow if possible, since the sides of a deep pit can collapse and bury the digger.

Of the three seepage pits shown here, the one of masonry block at left, below, may be the commonest; it is one of the most useful and is fairly simple to build, but professional help is generally needed to install its precast concrete cover.

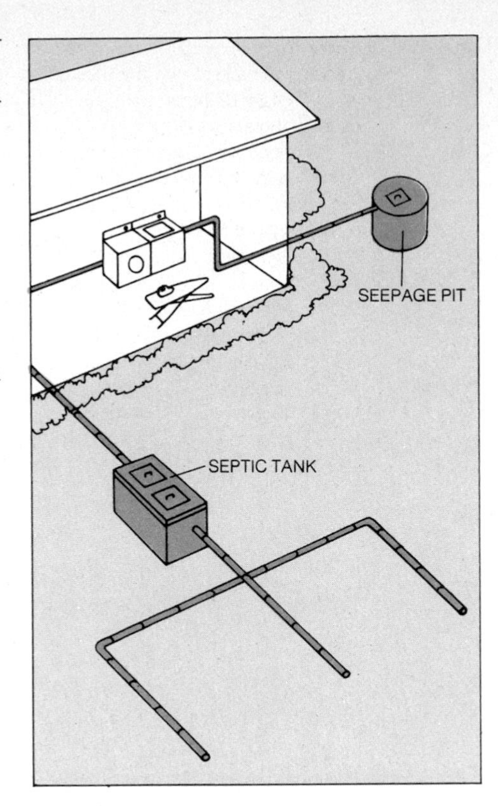

A place for clean waste. To handle the burden of waste water from a newly installed clothes washer, a seepage pit has been built a little to the side of this house, bypassing the septic tank. Such pits may also be connected to showers, lavatories, bathtubs and other fixtures that discharge uncontaminated waste water—but should not be used for toilets, kitchen sinks or dishwashers.

Three Kinds of Seepage Pits

A cylinder of unmortared blocks. This type of seepage pit—widely used and relatively easy to construct *(page 108)*—consists of hollow-core masonry blocks. The cores facilitate the passage of water out of the pit, but solid blocks or even ordinary bricks can be used instead, provided that ample space is left between them for the water to seep into the ground. The cover, of precast concrete, is best installed professionally, since it may weigh up to 500 pounds. The inspection hole in the concrete cover is optional; most seepage pits never require examination.

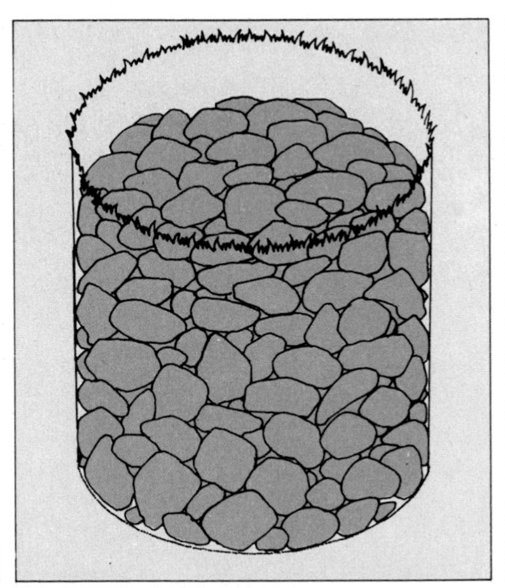

A stone-filled dry well. This simplest of all seepage pits—a hole completely filled with irregularly shaped rocks—is useful for absorbing small amounts of clean water, such as the runoff from an outdoor shower or a swimming pool. But its absorption rate is not sufficient to accept the large amount of water discharged by a clothes washer.

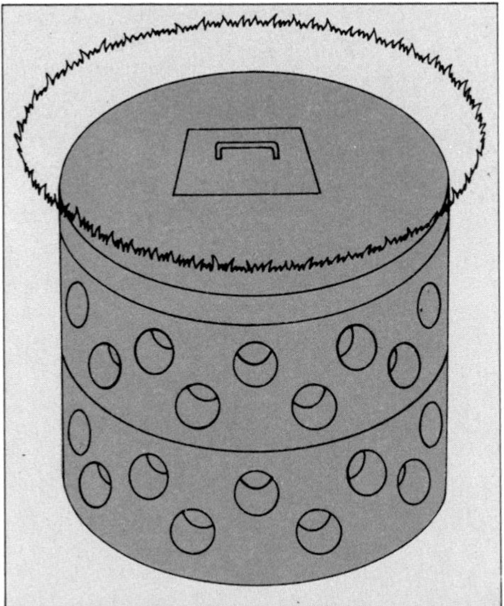

A drum of precast concrete rings. Pit liners like this one are sold as interlocking rings in diameters from 4 to 8 feet. Tapered openings allow water to flow out of the pit but help keep mud and gravel from seeping in. The rings, about 3 feet high and 3 to 4 inches thick, weigh over a half ton, and should be installed professionally.

Digging a test hole. To determine the suitability of the soil on the site you have chosen for a seepage pit, dig a test hole 4 to 12 inches wide and 2 feet deep, using a posthole auger *(right)* or an ordinary shovel. If you encounter clay or impenetrable rock, stop; fill up the hole and move to another site in your yard. If after digging 2 feet down you have encountered no obstructions, you can conclude that the soil itself is worth testing.

Place about 2 inches of gravel or small stones at the bottom of the hole. Then fill the hole with water and let the water seep away into the surrounding soil. Wait for several hours, or even overnight, and refill the hole with water. After the water has seeped away for the second time, the soil will be saturated—and ready for testing.

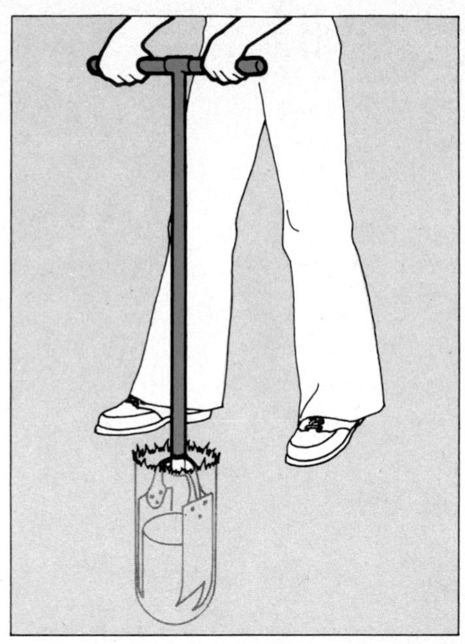

Measuring the percolation rate. To find out how fast your soil absorbs water, improvise a simple measuring device like the one shown here: a wooden dowel, with inches marked on it in pencil, and a wooden disk affixed to its end. Fill 6 inches of the test hole with water; place the device in the hole so the disk floats on the water's surface—and sinks as the water drops. Across the top of the hole, lay a board to serve as a fixed reference point, and record the number of minutes the water takes to drop 1 inch. Wait 30 minutes and repeat the test. Then wait and repeat again until you get the same result on two successive measurements. You then will have the rate of percolation—which may range from a few seconds to 30 minutes. Anything higher is unsuitable for a seepage pit. Use this figure in calculating how big a pit you will need.

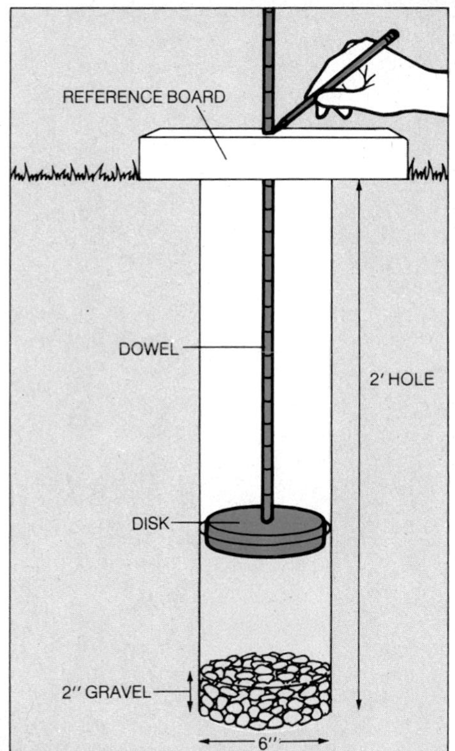

Tips for Pit Builders

No matter what kind of seepage pit you decide to build, there are some rules of thumb that apply to all.

☐ A seepage pit should be at least 100 feet from the nearest water well, 20 feet from any building and 10 feet from the lot line. The space between the house and the pit should be unobstructed.

☐ A seepage pit should be at least 2 feet above ground water and 5 feet above any impermeable bedrock. You can get valuable clues to the nature of the soil in your general area by studying the strata of nearby road cuts, stream embankments and building excavations. Another useful source is the soil maps of farm areas published by the U.S. Department of Agriculture.

☐ The best soil for drainage is sandy; it falls into clods that crumble easily. Soil that is dense, hard and dry, dull gray or mottled is apt to be nonabsorbent. Inspect the soil condition in your own yard by turning up spadefuls at several locations. When you have found a site that seems suitable, dig a test hole and make some experiments *(left)*.

☐ It is best to build on a side of a slope running down from the house, where gravity helps drainage. Do not build in a hollow where rain forms puddles.

☐ The average clothes washer—the appliance most commonly linked to a seepage pit—discharges 150 gallons of water per load. Design the pit for a capacity of 200 gallons to guard against overflow during heavy rains and winter freezes, when absorption is reduced.

Seepage Pit Dimensions

Percolation rate	Pit dimensions
0-5 min. per in.	4 ft. deep × 5 ft. diam.
6-10	4 ft. deep × 7
11-15	4 ft. deep × 9
16-20	5 ft. deep × 8.5

Fixing the proportions. A seepage pit can be deep and narrow, or shallow and broad—so long as the absorption area (the square footage of the pit's walls) is large enough to deal with the percolation rate of the soil in your own backyard. However, shallow pits are easier—and far safer—to dig: a pit over 5 feet deep calls for caution. The table at left contains data for pits with a capacity of 200 gallons. Detailed tables can be obtained from state and local boards of health.

Building a Pit with Concrete Blocks

Planning the pit. The drawings at right and below show the layout of any cylindrical concrete-block pit, whatever its size. Seen from above, the base of the pit consists of a footing of blocks laid radially—like wheel spokes *(right)*. The wall *(far right)* has courses of blocks set end to end—like sausage links—with the corners of the blocks touching each other on the inner circumference of the circle. Seen in cross section *(below)*, the blocks that form the wall stand crosswise to the blocks that form the footing. A pipe enters near the top; slanting slightly downward, it brings waste into the pit. A 6-inch-deep bed of gravel fills the bottom of the ring, and a 6-inch-thick sheath of gravel fills the spaces between the blocks and the surrounding earth.

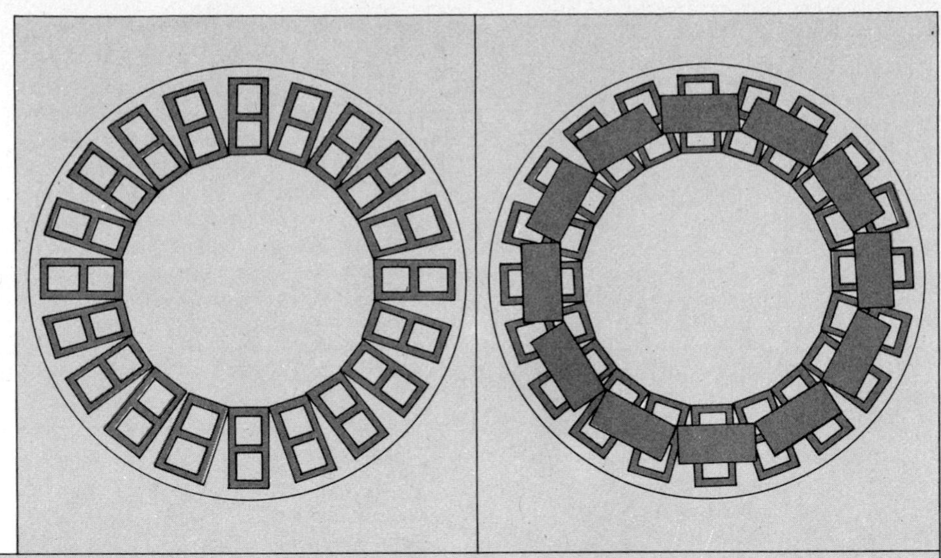

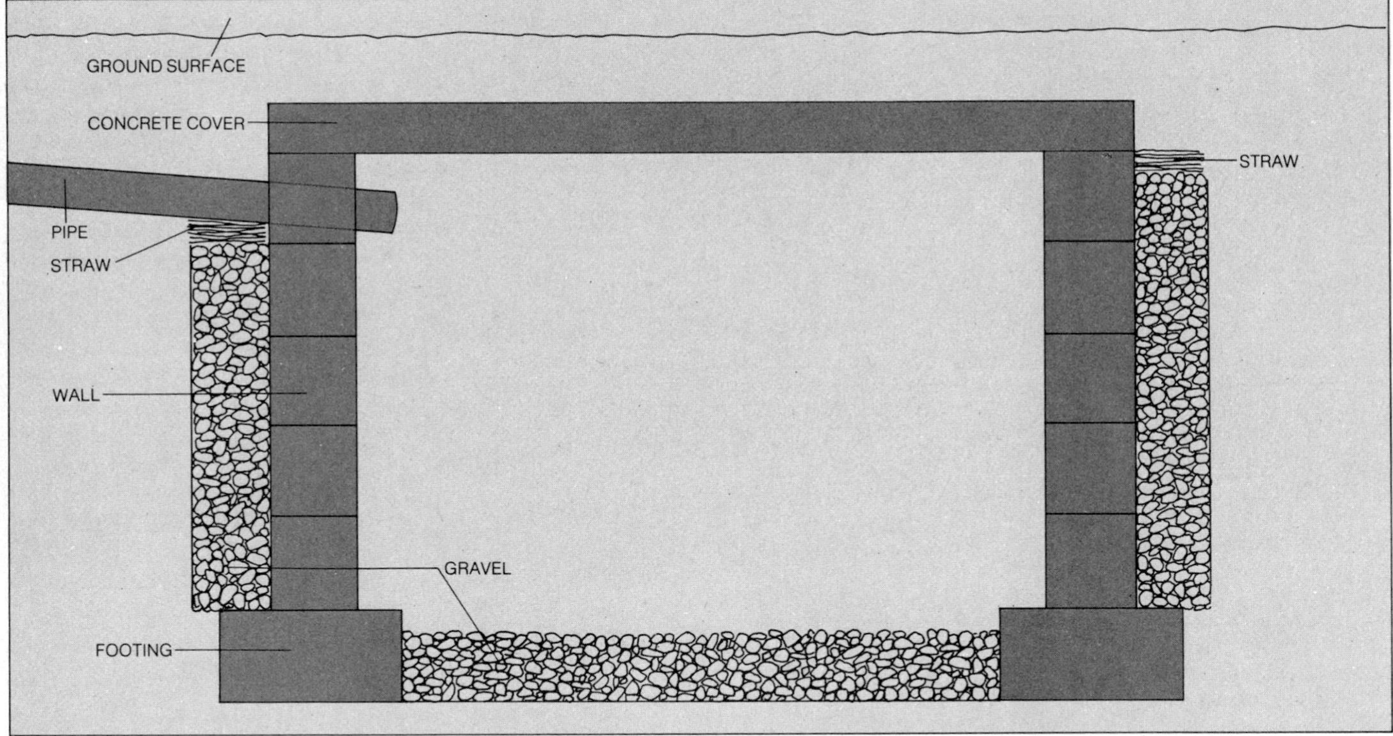

How Much Gravel, How Many Blocks?

To determine how much gravel you need for the 6-inch-thick bed of a pit, you must compute the area of the bottom. The mathematical formula for the area of a circle is pi (a number roughly equal to 3.14) times the square of the radius. (Use the table on page 107 to find the diameter of your own pit; its radius is one half the diameter.) In a simple example, the area of an 8-foot-wide pit is 3.14 times 4 times 4. For each square foot of bottom area, you need two 50-pound bags. To determine how much gravel you need for the wall filler, first calculate the pit circumference —it is 3.14 times the diameter of the pit —then multiply the circumference by the depth to get the wall area. Again, use two bags for each square foot.

To determine the masonry blocks needed for the pit footing, work out the footing's inner circumference in inches and divide this figure by eight. To figure how many you need for the walls, work out the circumference of the courses above the footing and divide by 16; this will give the number of blocks to a course. Multiply that figure by the number of courses—the pit's depth in inches, divided by eight.

1 **Building the lining.** Dig a circular hole approximately 2 feet deeper and 3 feet wider than the inner dimensions of the pit. Level the bottom and remove any loose dirt and rocks. To form the footing for the pit lining, lay a ring of blocks with each block arranged lengthwise toward the center and with the hollow cores facing upward. In the circle formed by the inner edges of the blocks, pour a bed of coarse gravel 6 inches deep.

To build up the wall, lay a row of blocks on top of the footing, this time with the blocks set end to end around the pit's circumference and with the hollow cores running horizontally. The blocks should touch along their inner perimeter, and their outer edges should lie about 6 inches from the edge of the hole. Stagger the second course so that each block rests athwart the two below it and, in succeeding courses, alternate the placement. Stop when you have come within 6 inches of ground level. No mortar is used to join blocks.

Fill the space between the pit lining and the side of the hole with gravel up to the flow line (the point at which the water enters the pit from the drain). On top of the gravel, place a 2-inch layer of straw to keep soil from sifting into the gravel.

2 **Connecting the drain.** Stake out a straight line from the point on the house foundation where the drainpipe exits to the point where it will enter the pit. Along that line, dig a trench about a foot wide and at least a foot deep. The trench must slope at least 1 inch for every 50 feet.

Using a star drill and a hammer, drill a 4-inch hole through the house foundation, as near as possible to the clothes washer, and high enough so the water coming from it will flow downward to the trench. Then assemble lengths of 4-inch pipe (plastic is generally used) in the trench. Insert one end through the hole in the house foundation and the other through the pit lining so that the pipe protrudes about 6 inches into the pit. The pipe end can be passed through a hollow core in the topmost course of blocks, or one block can be removed to make room for it.

3 **Covering the pit.** Have a building-supply firm deliver and install a concrete pit cover that fits over the outer diameter of the pit lining. With the cover securely in place, fill the remainder of the hole and the drain trench with soil to grade, and tamp down firmly. There should be at least 6 inches of soil above the pit cover and the trench.

Assemble the clothes washer as you would a sewer-connected unit *(page 104)*, then connect the drain to the pipe that goes to the seepage pit. Run the washer through a complete cycle to be sure the drain works properly.

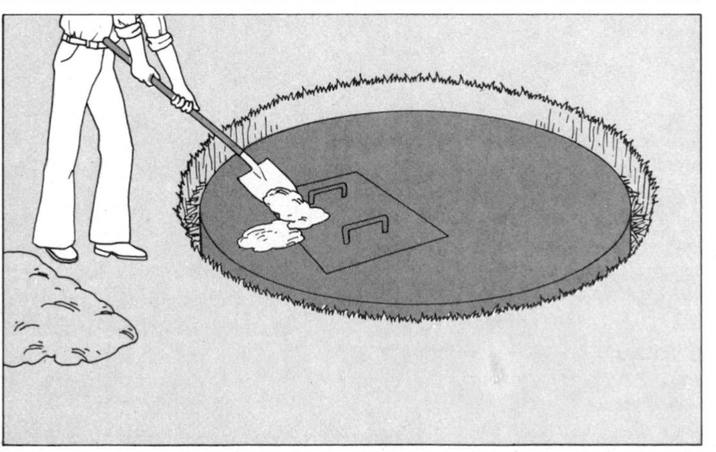

Making Life Easier with Outdoor Plumbing

A 100-foot garden hose lies in an unsightly pile in the backyard; to use it, you must first untangle it, then drag it like a reluctant serpent around the corners of the house and across the lawn. Children and grownups come in from play on a hot afternoon and track muddy footprints across clean carpets from the back door to an upstairs shower. Such annoyances may be avoided simply by extending water supply lines outside the house to serve you where you need them, and then fitting them with specialized attachments. Some of the most popular are built-in sprinklers for a lawn and the freezeproof sillcock, which has an outdoor spout and handle connected by a long shaft to a valve inside the house, where it cannot freeze up.

The best place to tap a water supply for an outdoor extension is generally the service line in a basement *(page 113)*; if that line is not accessible, you can use an existing standard sillcock *(opposite)*. Do not use the freezeproof type for this purpose, however—it will not accommodate the necessary fittings. The pipe for an extension must be buried underground.

In most regions, outdoor pipes must be drained in the fall to avoid winter freezing and bursting. If you can, slope the pipe toward the house so that the water drains from a valve in the basement *(page 113)*. If your house has no basement or your lawn slopes away from the house, pitch the pipe down from the house so that the water drains into a gravel pit through a device, called an automatic drain valve, attached to the far end of the buried pipe *(pages 112-113)*. Drainpipes for showers *(pages 114-117)* must also be pitched and buried.

The most convenient piping for most outdoor extensions is polyethylene (PE); check your local code, however, to see if there are special requirements for making the indoor connection to the service line. Polyethylene can be joined with simple clamps and is flexible enough to take gentle curves; it does not require many fittings, an important point, since each fitting tends to reduce water pressure at the outlet. Where a large number of fittings is required—in a lawn sprinkler system, for example—polyvinylchloride (PVC) is preferable to polyethylene; it is less flexible but is joined with strong cement fittings.

The size of the pipe you use depends on the size of your water source. If the existing supply pipe is ¾ inch or larger, use ¾-inch pipe; if it is less than ¾ inch, match the size to the existing pipe.

Installing a Sillcock

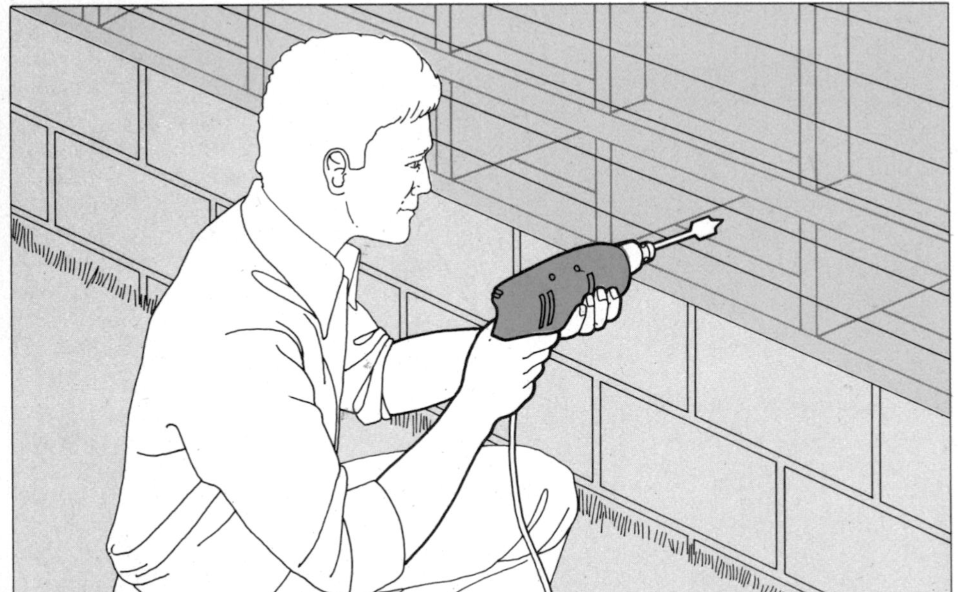

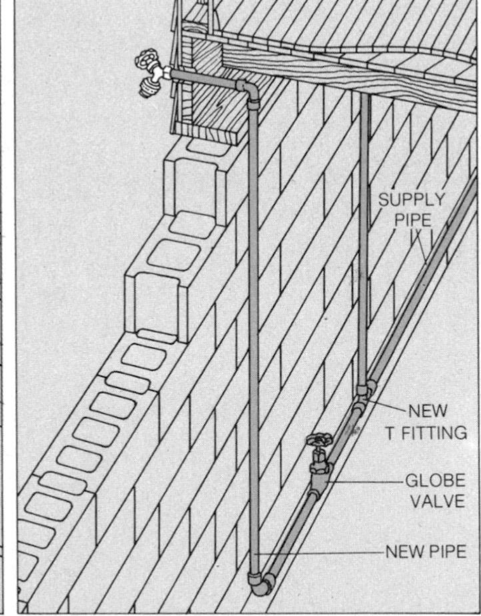

1 Boring a hole and choosing a sillcock. If possible, locate the sillcock on the wood wall of the house, just above the masonry foundation; make sure that the corresponding interior surface of the wall is clear of obstructions. Using a spade bit and, if necessary, an extension, bore a hole through the wall just large enough for the pipe. If you must make a hole in masonry, use a star drill, setting the point on a horizontal strip of mortar or at the center of a cinder block. Drive the drill with a ball-peen hammer, and turn the point a few degrees between blows. Attach the sillcock to a length of pipe 3 inches longer than the thickness of the wall, using an adapter if necessary. A freezeproof sillcock *(page 36)* does not require a separate pipe: select a sillcock with a stem at least 2 inches longer than the wall thickness, and angle the hole slightly downward toward the outside.

Whatever type of sillcock you use, fasten it in place now. Insert the assembly in the hole you have bored, and screw it to the wall, using anchors if necessary. If the flange of the sillcock does not lie flush with the outside wall, fill the gap with waterproof silicone caulking.

2 Tapping the water supply. Install a T fitting *(page 71)* in a convenient supply pipe, run a short section of pipe and install a globe shutoff valve at the end of it. Install an elbow on the end of the sillcock pipe; if necessary, use an adapter to connect the elbow to the stem of a freezeproof sillcock. Then run pipe from the valve to the elbow. Finish the job by filling the hole around the pipe inside the basement with waterproof silicone caulking; use a putty knife to smooth and feather the patch.

Running Pipe from an Existing Sillcock

1 Installing a T fitting. If the sillcock is on threaded pipe *(top right)*, remove the screw that fastens the sillcock to the wall and, using a 10-inch pipe wrench, unscrew the sillcock from the pipe. Screw a T fitting with a tapped inlet onto the pipe, with the inlet pointed down. Screw a nipple to the open arm of the T and screw the sillcock to the nipple; screw a plastic adapter to the inlet.

If the sillcock is on copper pipe *(right center)*, locate this pipe inside the wall and cut it at least 2 inches from the wall. Remove the screws securing the sillcock to the outside wall and pull the sillcock and the length of pipe attached to it completely out of the wall. Using a propane torch, separate the section of pipe from the sillcock *(page 75)* and sweat a T fitting with a tapped inlet onto the pipe section, followed by a spacer and the old sillcock; the inlet should point straight down when the sillcock is upright. Push the new sillcock assembly back into the hole and, inside the basement, rejoin the two sections of pipe with a coupling. Screw a plastic adapter to the T inlet.

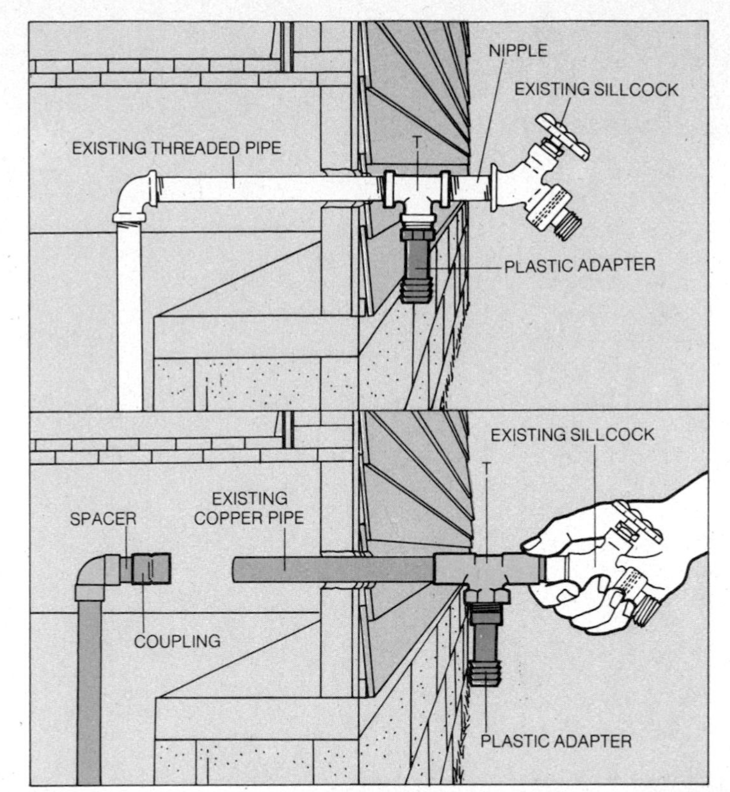

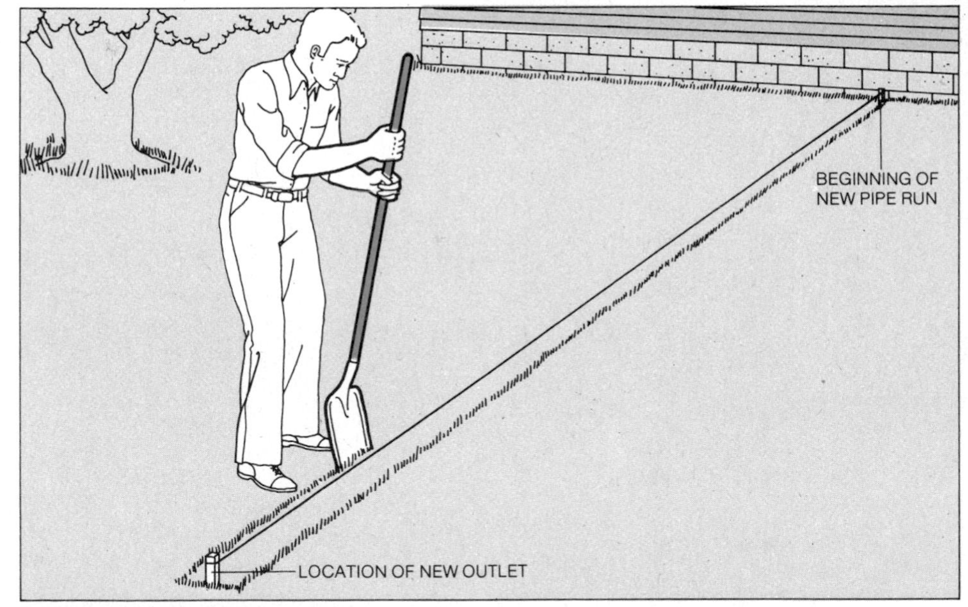

2 Staking out a trench. Drive a stake into the ground at the point you have chosen for a new outlet, and another stake at the point on the house from which you will run the new pipe; tie a string between the two stakes. The trench for the pipe will be about 6 inches wide. To lay it out, make two parallel grooves in the ground, about 3 inches to the left and right of the string, by repeatedly pushing the blade of a flat shovel about 2 inches into the ground. Remove the string and stakes, and connect the two grooves at their ends to form a rectangle. Before lifting off a sod cover, divide the rectangle with your shovel into segments no more than 5 or 6 feet long and spread a plastic strip along the ground next to one of the grooves. Push the shovel under the sod at the corner of a segment and work it up and down to free the roots. Repeat along the rest of the segment, then lift the sod carpet and place it on the plastic sheet.

3 Digging out the soil. Spread a large plastic sheet or a tarpaulin on the ground. Working in from the sides of the grooved rectangle, dig a V-shaped trench about 10 inches deep. Pile the loose soil on the plastic sheet.

4 **Tunneling under a paved walk.** Have a 3-foot length of ¾-inch steel pipe threaded at one end, and buy an adapter that will fit the threaded pipe to the threaded end of a garden hose. Turn the water on full force and stick the end of the steel pipe into the ground under the walk, using the pipe as a water-pressure pick to loosen the earth. The stream of water should be strong enough to keep the open end of the pipe clear and to sweep away loose earth. When you reach the middle of the walk, cross to the other side and work from there to complete the tunnel.

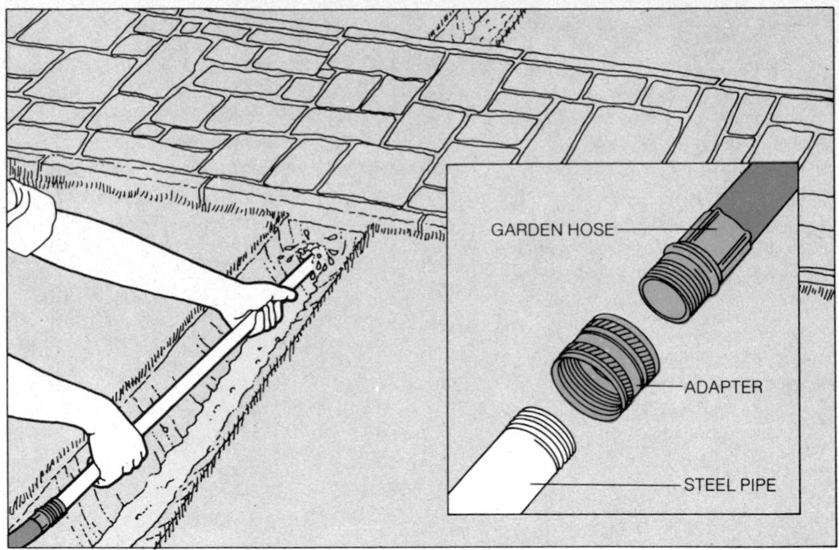

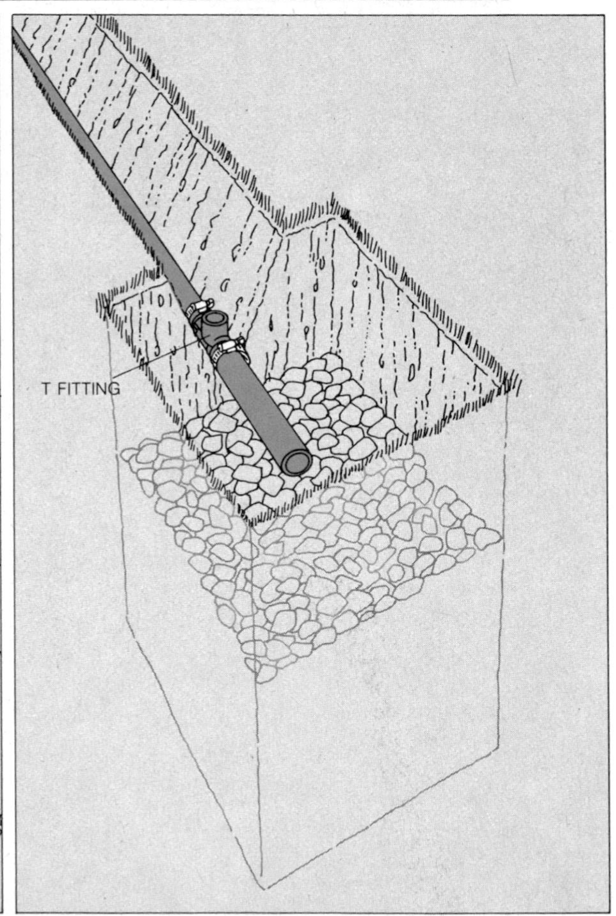

5 **Laying and pitching pipe for drainage.** Run pipe along the complete length of the trench. Then, beginning at the far end, prop up the pipe with fragments of brick or stone at 6-foot intervals to make it slope gently downward from the house. To check the pitch, set a carpenter's level on the pipe at several points along its length—the bubble in the level should always be slightly to the house side of center. Shore up the pipe between the brick fragments with loose earth.

6 **Digging a drainage pit.** At the far end of the trench, dig a pit 1 foot square and about 20 inches deep. Fill the pit with coarse (1-inch) gravel to the level of the trench bottom. At the end of the pipe in the trench, directly beneath the location of the new outlet, install a T fitting with the inlet pointing straight up. Add about 6 inches of pipe to the open arm of the T, so that the end of the pipe now reaches the middle of the drainage pit.

7 Installing an automatic drain valve. Screw a threaded adapter to the open end of the pipe and an automatic drain valve, or ball drip, to the adapter. The automatic valve consists of a chamber containing a loose metal ball *(inset)*. When water flows through the pipe, its pressure forces the ball against a drain hole, and water flows to the outlet aboveground. When the water is turned off and the pressure drops, the ball rolls back into the chamber and the water remaining in the pipes flows out through the drain hole.

Cut a piece of galvanized steel pipe slightly larger in diameter than the drain valve and twice as long, and slip it over the valve; the pipe will help keep the drain hole free of blockage. Add another 3 or 4 inches of gravel to the pit.

8 Connecting the pipe to the sillcock. Install an elbow on the house end of the pipe below the sillcock. Measure the distance from the elbow to the adapter placed in the sillcock T in Step 1 and install a length of pipe between the two. Apply two coats of latex paint to the section of pipe above ground—both flexible and rigid plastic pipe deteriorate in direct sunlight.

Install risers in the T for the new outlet at the far end of the pipe *(pages 114 and 121)*, then fill in the trench. After the outlet connections are made, shovel earth into the trench and the pit to about 2 inches above the pipe and hose it down. Next day, tamp the earth with a 2-by-4, fill in the trench and pit, and, if you have removed sod cover, replace it.

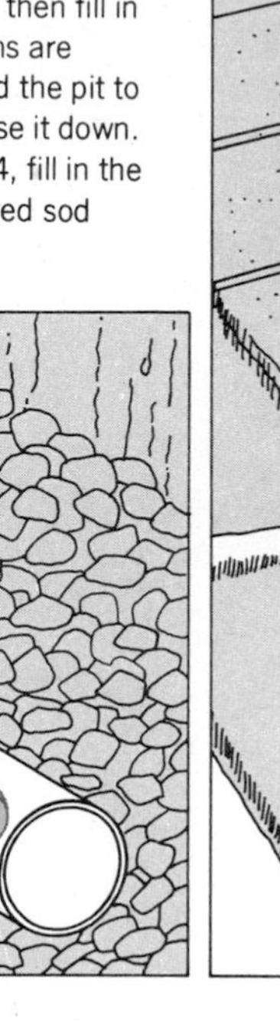

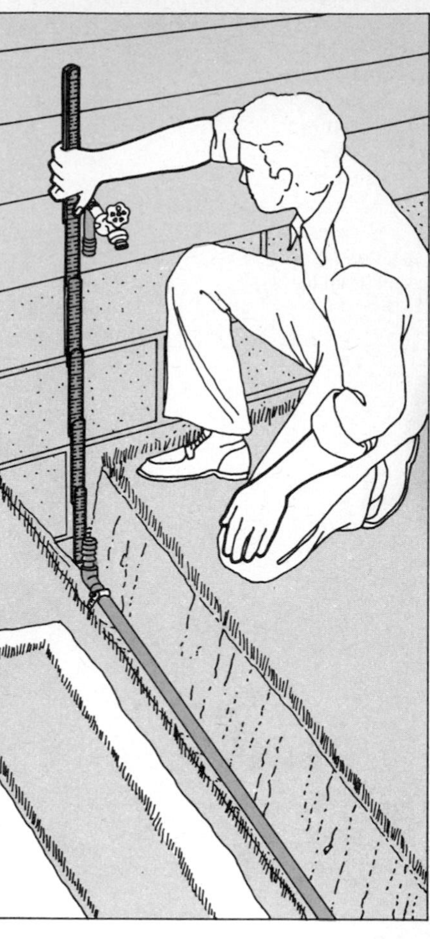

Tapping and Draining Water in a Basement

A valve-and-drain combination. If the ground outside your house is level or slopes slightly downward toward the house, you can tap water for a distant outlet from a basement supply line and drain the system without a drainage pit. Dig a trench for outdoor pipe *(pages 111-112)* and bore a hole through the house wall at the bottom of the trench *(page 110, Step 1)*; pitching the hole slightly downward into the basement.

In the basement, install a T-fitting in the service line *(page 110, Step 2)*, followed by a short length of pipe for a spacer. To this spacer attach a stop-and-waste valve—similar to a globe shutoff valve, but with a tap to drain water from the run of pipe beyond it. Use adapters if necessary to accommodate different pipe and fitting materials. Be sure that the arrow on the valve points in the direction of the flow of water. Run pipe from the valve through the wall and to the end of the trench, pitching the pipe downward toward the house. At the far end of the pipe, install an elbow for the fixture connection.

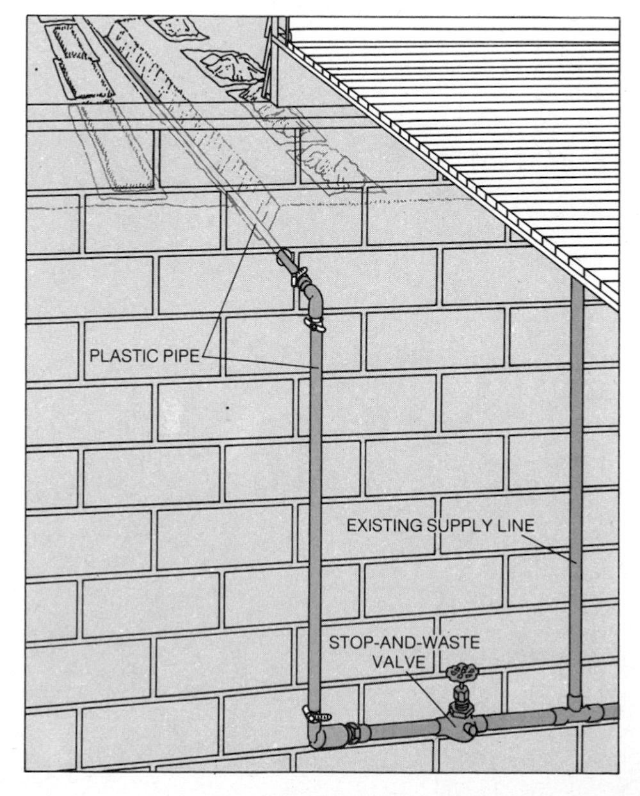

Bracing the hydrant. Cut a 40-inch-long support for the hydrant from ¾-inch steel pipe. Set one end of this pipe next to the end of the pipe in the trench and drive the support pipe about 2 feet deep into the earth with a sledge hammer.

For the vertical riser of the hydrant you will need a brass pipe, threaded at both ends and about 8 inches longer than the height of the hydrant above the ground. (Brass pipe is expensive, but plastic is not strong enough for this job and galvanized steel corrodes too quickly.) Attach this pipe to an adapter that connects to the T or elbow at the end of the underground pipe. Strap the two vertical pipes together with two stainless-steel pipe clamps. Turn on the water briefly to flush dirt from the pipes, then screw the hydrant to the brass pipe and turn the water on again to test the system for leaks.

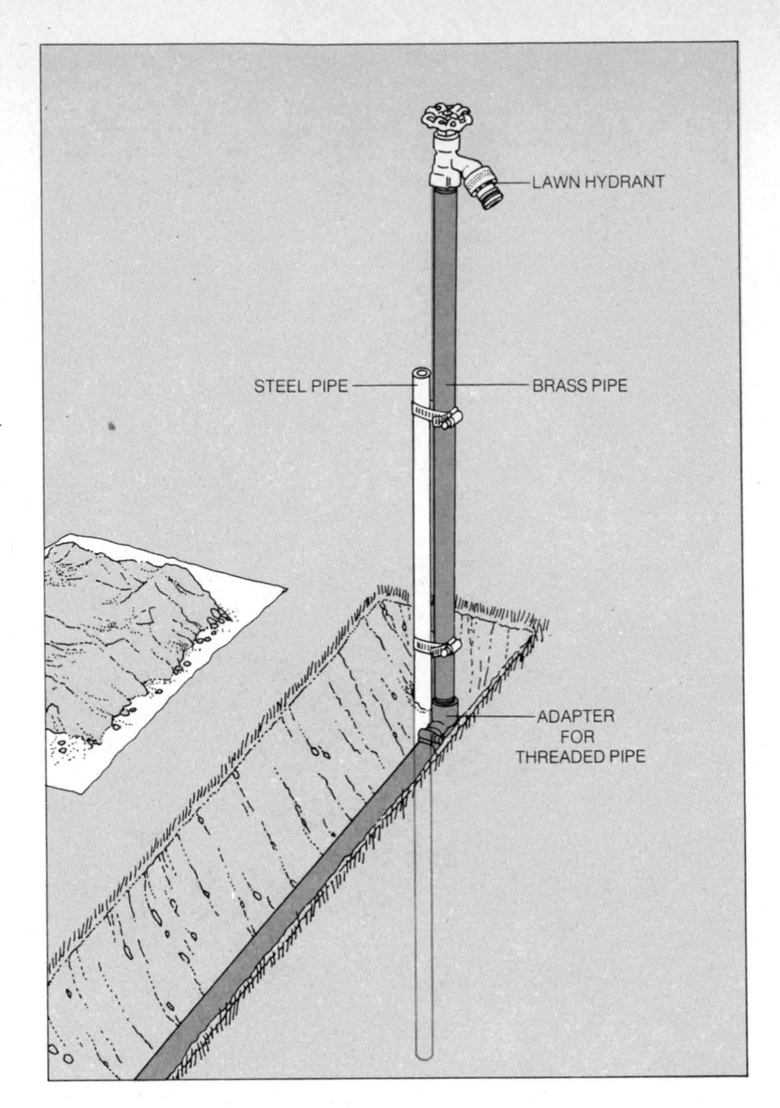

A Shower on an Outside Wall

1 **Digging a foundation pit and a dry well.** Dig a foundation pit 1 foot deep and about 6 inches larger all around than the shower base. Dig a dry well 3 feet square and 4 feet deep downhill from the shower and at least 10 feet away from both the house and any water well on your property. Connect the two pits with a trench for outdoor piping *(pages 111-112)*.

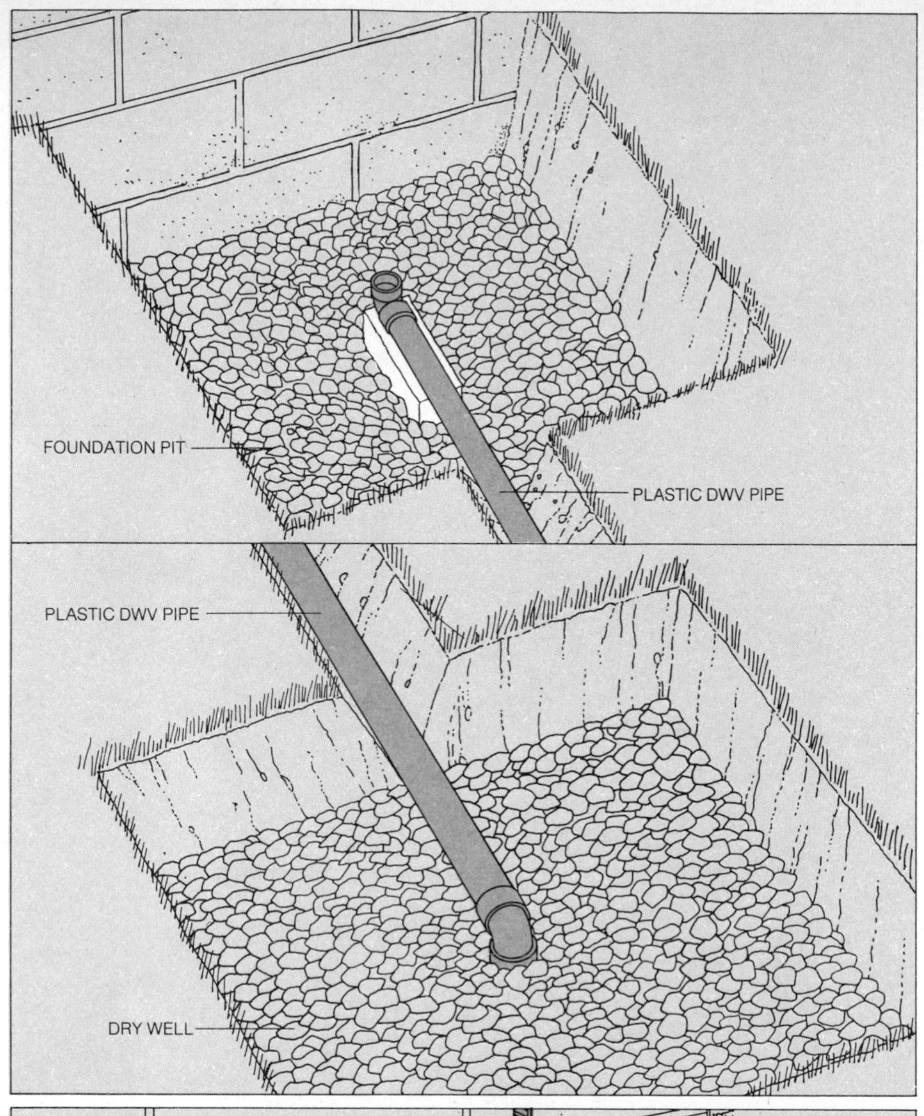

2 Installing the drainpipe. Fill the well and the foundation pit with 1-inch gravel to the level of the trench bottom. Mark the location of the center of the shower-base drain with a stake. Run plastic DWV pipe of the same diameter as the shower-base drain from the stake to the center of the dry well; pitch the pipe to drain into the well.

Fit a 90° elbow to the end of the pipe in the foundation pit; point the elbow directly upward and set its open end at the point marked by the stake. Then fit a 90° elbow to the end of the pipe in the dry well; point this elbow downward, and dig a small hollow in the gravel for its open end.

3 Installing the vertical drainpipe. Place a straight piece of wood across the top of the gravel pit and measure the distance from the bottom of the elbow lip to the bottom of the wood. Remove the drain strainer from the shower base and measure the depth of the drain hole. Add the two measurements, subtract ½ inch and cut a length of drainpipe equal to the total. (In a typical installation, the distance from the elbow to the top of the pit will be 4 inches, the depth of the shower drain 1½ inches, and the length of the pipe 5 inches.) Install the pipe in the elbow.

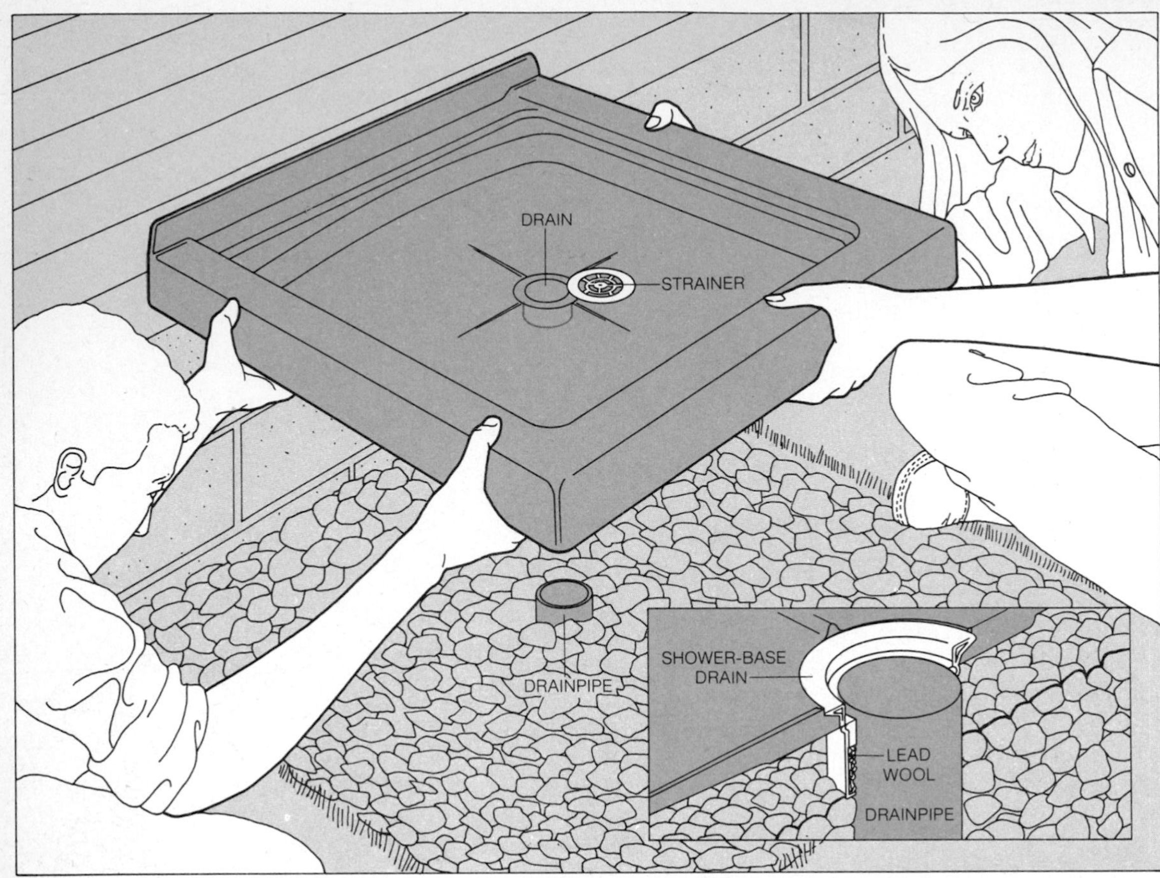

4 Installing the shower base. Fill the trench with loose earth; if you have removed a sod cover, replace it *(page 111)*. Fill the foundation pit with gravel to ground level—do not let any gravel drop into the drainpipe—and tamp the gravel down with a 2-by-4. With a helper, lower the shower base carefully onto the exposed end of the drainpipe. When the base is firmly in place, pack the space around the drainpipe tightly with lead wool *(inset)*. Replace the strainer plate.

Fill the dry well with gravel to about 4 inches below ground level in successive 1-foot layers, tamping tightly as you add each layer. Then fill the well with loose earth to ground level and, if you have removed a sod cover, replace it.

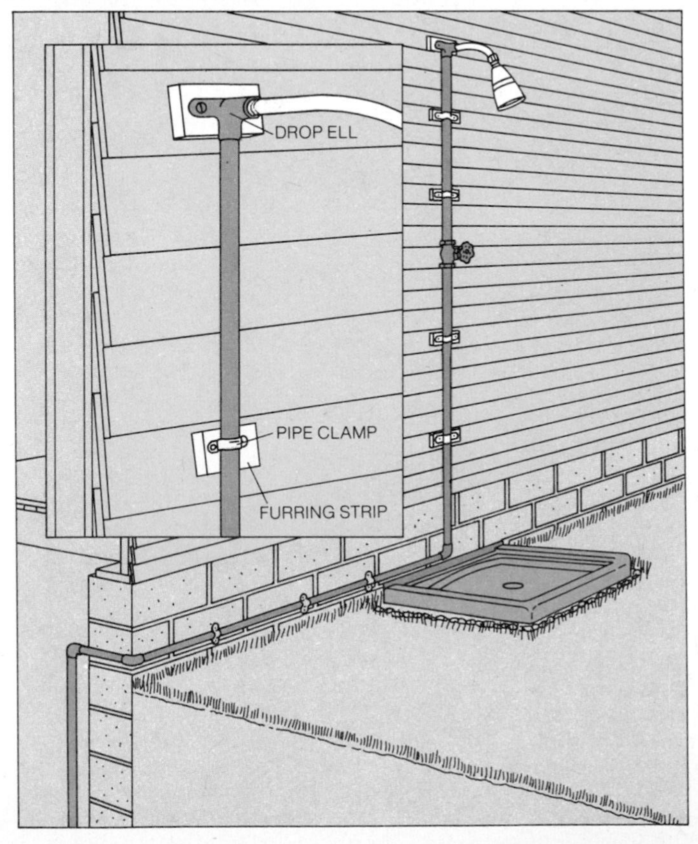

5 Bringing water to the shower. Run a water supply pipe along the house wall to a point directly above the center of the shower base, then straight up about 4 feet. Install a shutoff valve at this point, then add about 2½ feet of vertical pipe topped by a drop ell with a tapped inlet. Fasten the drop ell to the wall with screws, and clamp the pipe to the outside wall. On a house with wood siding *(inset)*, clamp the vertical pipe runs to 3-inch-long blocks nailed to the side of the house at 2-foot intervals. Screw a shower arm and a shower head into the drop ell.

A Freestanding Shower

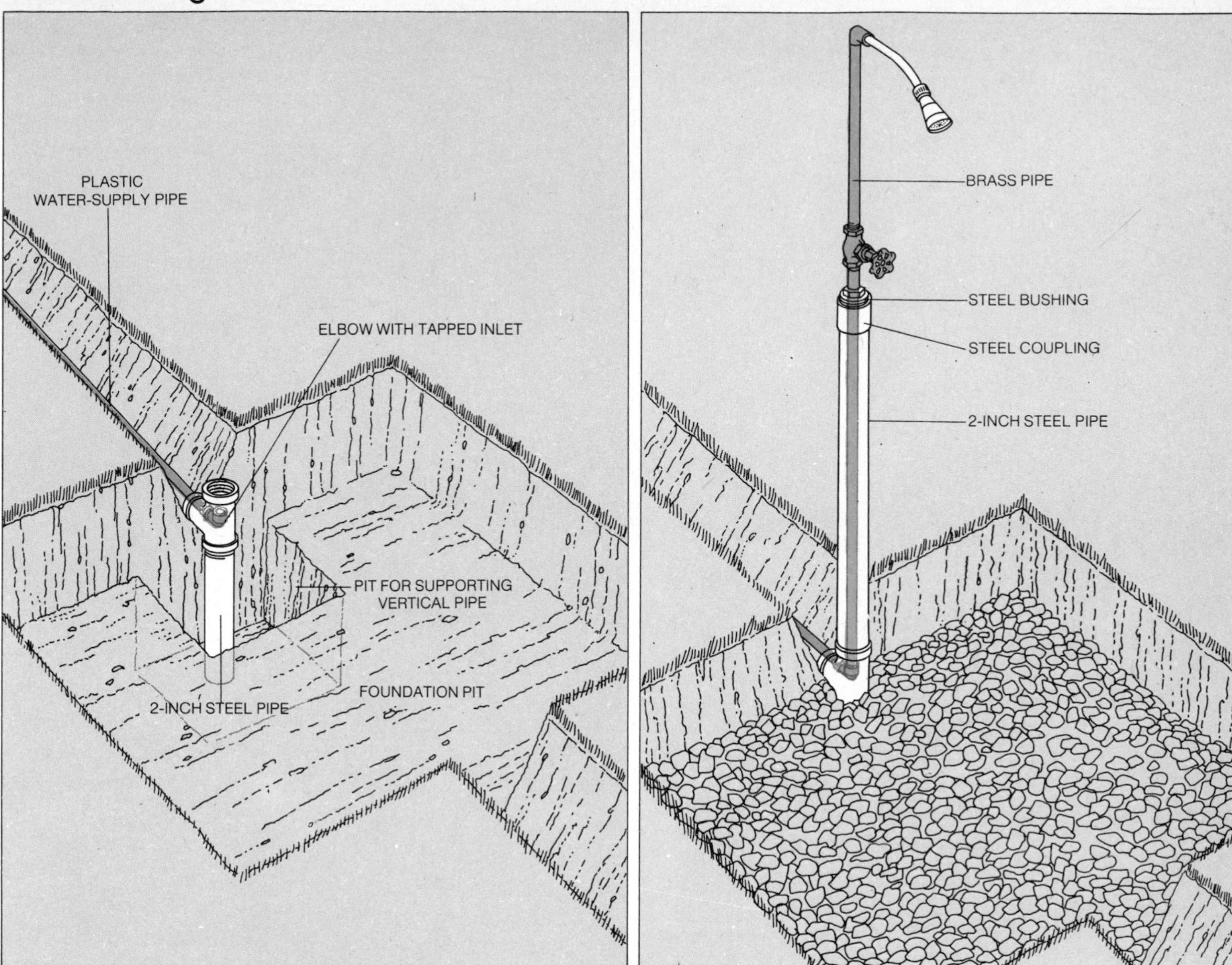

1 Supporting the vertical supply pipe. Dig a foundation pit and a dry well *(page 114, Step 1)*. Inside the edge of the foundation pit nearest the house, dig another pit about 1 foot square and 1 foot deep. Dig a trench for supply pipe from the house to the foundation pit and a trench for drainpipe from the foundation pit to the dry well *(pages 111-112, Steps 2 through 4)*.

For the next part of the job, use a piece of 2-inch galvanized steel pipe, threaded at one end and long enough to reach from the bottom of the 1-foot pit to the bottom of the trench. Screw a steel T fitting to this pipe and set the pipe in the middle of the pit with the threaded end up and the open inlet of the T pointing into the trench.

Run supply pipe from the house to the foundation pit, add an elbow with a tapped inlet and push the supply pipe into the T with the inlet pointing upward. Prop the steel pipe upright and fill the foundation pit with 1-inch gravel to the trench bottom; tamp the gravel down with a 2-by-4.

2 Running pipe to the shower head. For this part of the job you will need a 4-foot length of brass pipe and a 3½-foot-long section of 2-inch steel pipe, threaded at both ends. Screw the brass pipe into the elbow at the end of the supply pipe, then slip the steel pipe over the brass pipe and screw it into the open arm of the T fitting. Screw a 2-inch steel coupling onto the top of the steel pipe and add a steel bushing just large enough to hold the brass pipe. Install a shutoff valve at the end of the brass pipe and top the pipe with a 3-foot-long section of brass pipe, a 90° elbow, a shower arm and a shower head. Complete the job with Steps 2 through 4, pages 115-116.

An Underground Sprinkler System

Few gardeners enjoy dragging a sprinkler around the yard to keep the lawn green during summer dry spells. No one has to: a sprinkler system buried beneath a lawn or garden can take over the job. At the turn of a handle or—in an automatic system—the timed flick of a switch, sprinkler heads push up through the grass and flowers to create overlapping umbrellas of spray. At the end of the watering period the heads sink into the ground.

Installing such a sprinkler system calls for no greater skill than does installing the lawn hydrant on page 114. The steps in both jobs are nearly identical—tapping a supply line, digging trenches, running pipe (normally, rigid plastic pipe for maximum pressure and ease of handling) and installing outlets. However, the sprinkler system must be controlled near the house, usually by a manual shutoff valve.

A simple lawn may have all its sprinkler heads on one supply line, controlled by a single shutoff valve. A yard divided into several distinct areas with different watering needs calls for a more complex system with several supply lines, each serving a single watering zone and fitted with a separate shutoff valve. The drawing at right shows such a system at work in the three most common types of watering zones: a sunlit front lawn that must be watered long and often; a shady area that needs briefer, intermittent watering; and a garden that is watered through sprinklers with high heads.

The so-called pop-up sprinkler heads that you are most likely to use are installed just below the surface of the ground, out of sight—and out of the way of lawn mowers. When the shutoff valve is turned on, water pressure forces an internal piston upward to just above ground level, and the water sprays in a full or partial circle; when the water is turned off, gravity or a spring within the head brings the piston down again. Smaller areas and garden beds can be covered by the simplest type of pop-up, consisting of a fixed piston whose only motion is up and down. An area greater than about 60 feet across must be watered by a more expensive rotary pop-up, which has a rotating piston or spout that shoots out long jets of water.

Designing a system that takes into account all the variables of site and vegetation is a job for an expert. Fortunately, it is also a job that an expert will do for you free of charge. Most distributors and all manufacturers of sprinkler parts will supply you with a blueprint of a system tailored to your home, in exchange for your promise to buy their parts.

To help the designer, you must supply him with certain information—a rough scale map of your house and grounds, together with data on plantings, plumbing system, soil and climate. The designer will have to know where you have grass and where you have flowers or bushes; which areas are sunlit and which shady; whether your soil is sandy, rocky or compact. If you use a public water system, he will need the internal diameter of your water meter in inches (the figure is usually stamped on the meter housing). For a private system he must have the pump discharge pressure and pumping height. For either system he should know the available water pressure in your supply lines. Your distributor will lend you a pressure gauge to get the last figure; use it by the method shown on page 120. Finally, indicate the diameter of the pipe from which you plan to tap water.

The blueprint you receive will indicate the type of sprinkler head to use in each zone, the exact location and settings (for full or partial circles) of the heads, and the layout and diameter of the interconnecting web of plastic pipes. This blueprint will not exactly match the plan at right; but its main elements will be much the same. Because these elements are common to all sprinkler systems, you can use the plan and the instructions as a guide for installing your own.

In the first stage of installation, the water supply is tapped and controls are installed. For manual controls, tap at a sillcock *(page 120)*, and for an automatic system, install the controls in the basement and tap water from an inside supply pipe *(page 123)*. Then dig the trenches, pitched away from or toward the house *(pages 112-113)*; run pipe and install the sprinkler heads *(page 121)*. Be sure to test the system before you bury it underground, so that if adjustments are needed *(page 122)*, you can make them with least trouble.

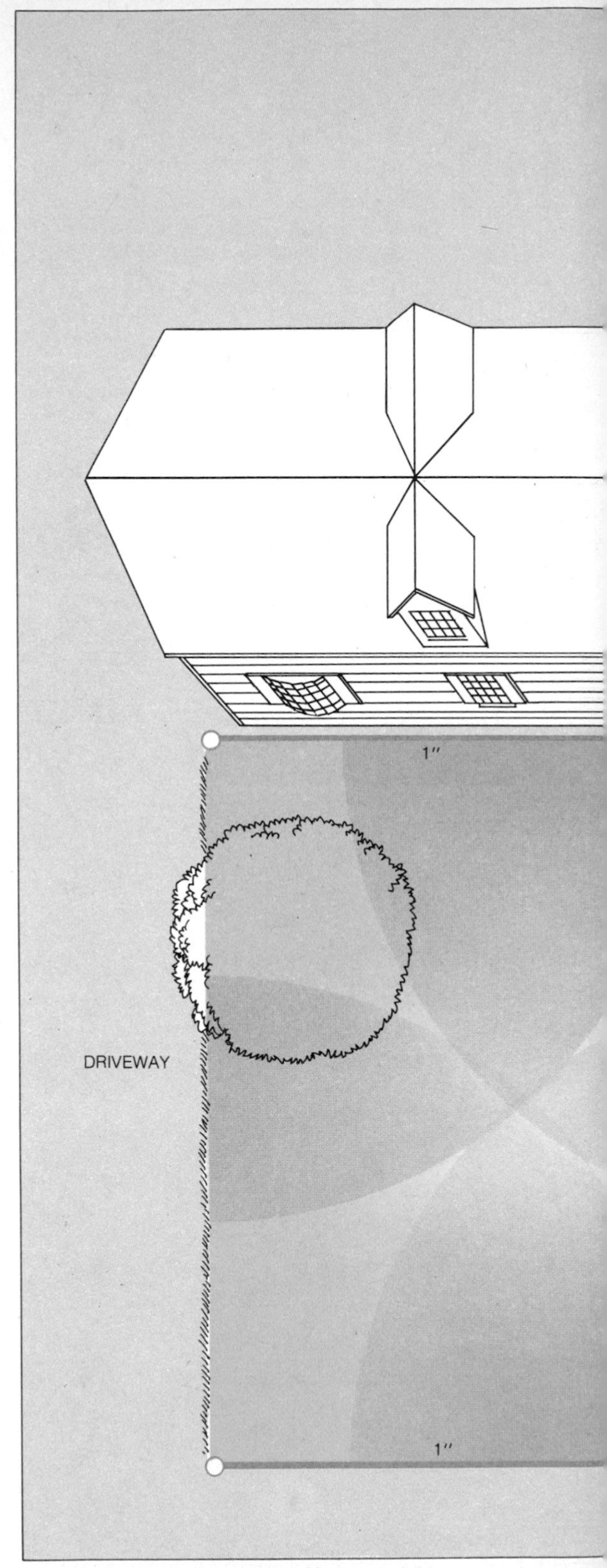

A three-zone sprinkler system. No two sprinkler systems are identical, but the one shown above illustrates types of problems and solutions that will appear in almost any system. It has three distinct watering zones: a sunny front lawn, a tree-shaded side lawn and a flower garden. Separate

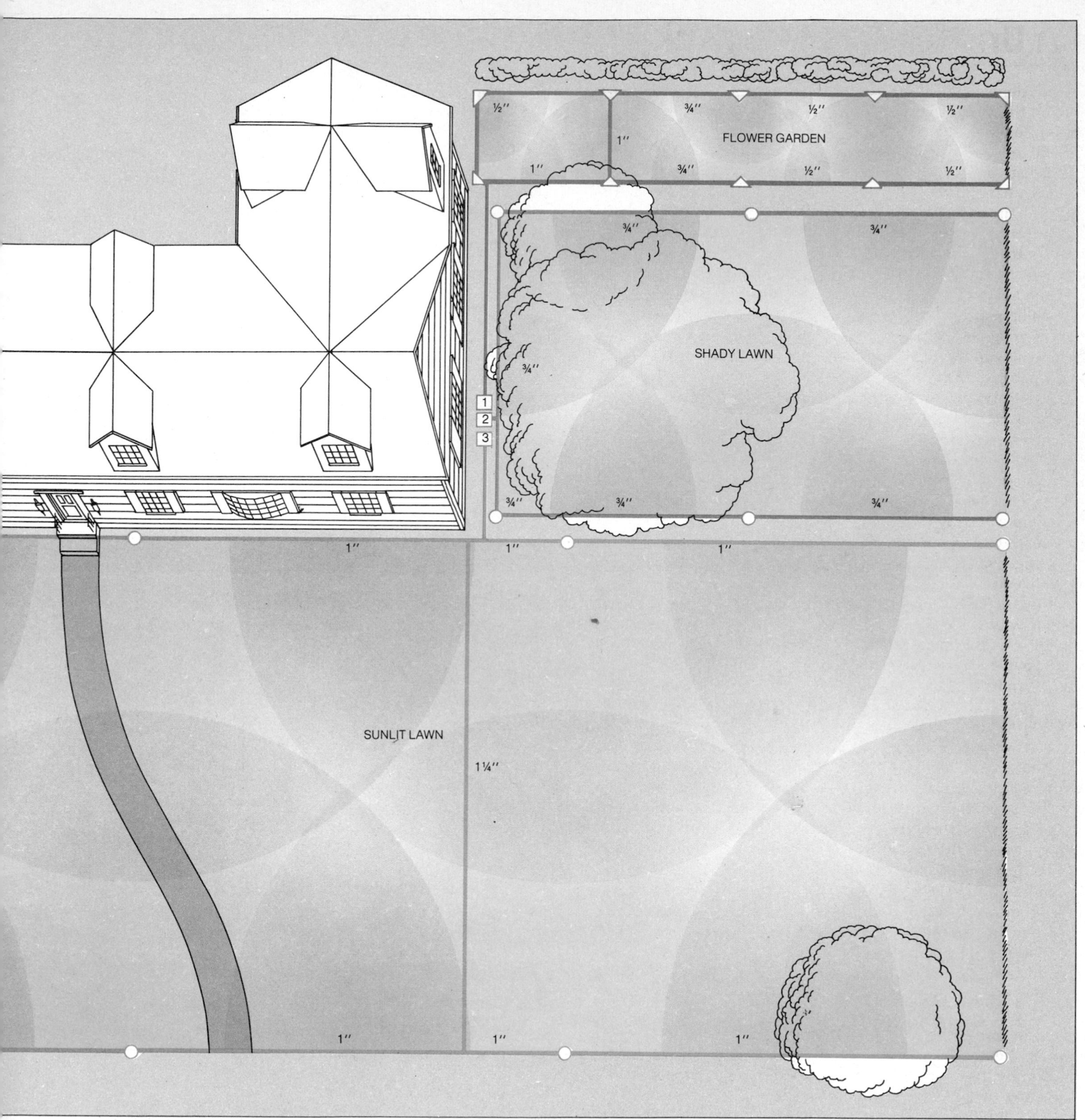

supply lines feed the three networks of pipe, called circuits, that serve the watering zones; shutoff valves (numbered squares) for all three circuits are clustered near the main source of supply. The layout of the pipes between the shutoff valves and the sprinkler heads is shown in green, yellow and blue, with the size of pipe indicated for each run. The flower garden at the top of the plan is watered by simple pop-up sprinkler heads (triangles); the two lawns, by rotary pop-ups (circles). When in operation, these heads spray water in the patterns indicated by large half and quarter circles. Each pattern overlaps other patterns by about 60 per cent, to allow for small imperfections in the layout of the heads, and to provide a more uniform depth of coverage—the outer rim of a head's area of coverage gets less water than the ground closest to the head.

Measuring the Water Pressure

Screw a pressure gauge (available from the supplier of your sprinkler parts) into a sillcock hose bib, using a hose adapter. If you do not have a sillcock, close the shutoff valve on a cold-water line below a fixture, disconnect the line, and screw the adapter and gauge onto the valve outlet. When no major appliance, such as a clothes washer, is in use, turn the sillcock or valve full on; the gauge will indicate the water pressure at that moment. On three successive days take readings at 7 a.m. and 4 p.m. (the times when most lawns should be watered). Average all six readings to determine your available water pressure.

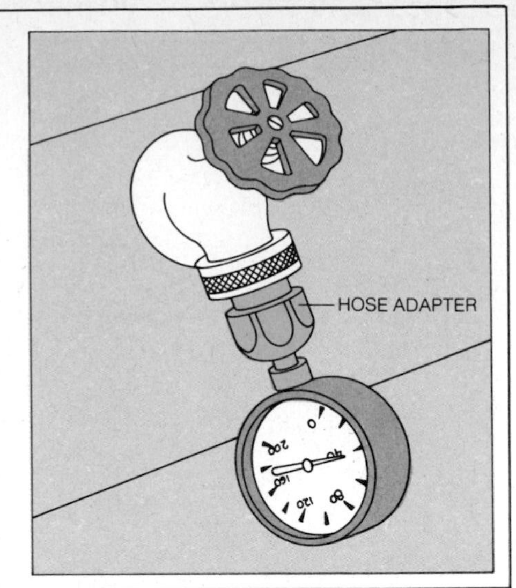

Controls for a Manual System

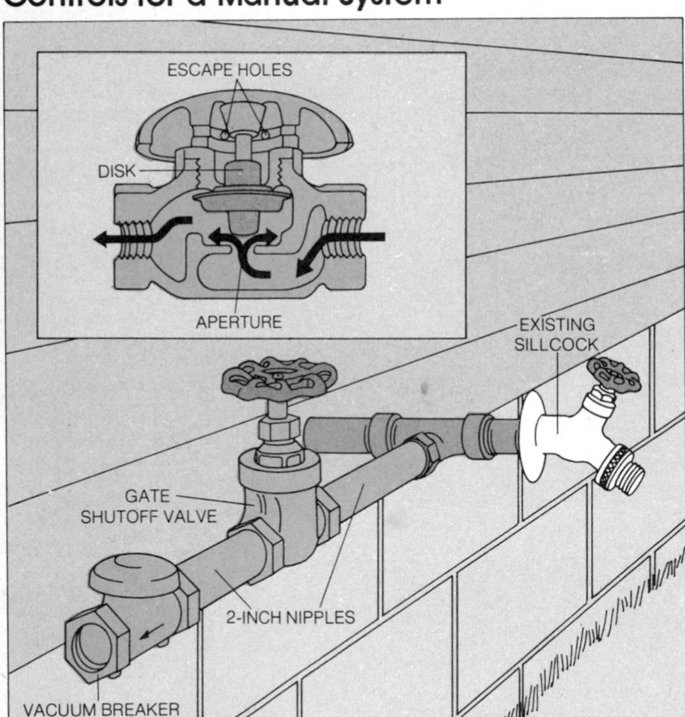

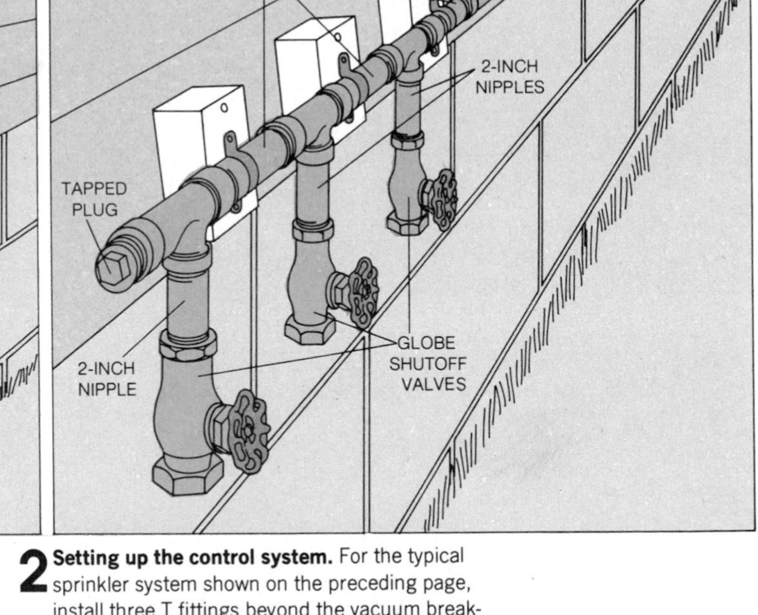

1 Tapping a sillcock. Install a tapped T fitting in a sillcock line *(page 111)*, with the inlet of the T set horizontally. Screw a 2-inch brass nipple into the T, a gate shutoff valve onto the nipple and a nipple into the valve. Then screw an antisiphon device called a vacuum breaker onto the nipple, with the breaker cap pointing upward and the arrow on the side of the breaker pointing in the direction of water flow. A raised disk inside the vacuum breaker *(inset)* permits water from the supply line to flow through an aperture and on to the sprinklers *(arrows)*. If water starts to flow backward, suction pulls the disk down, blocking the aperture; this water flows out of the breaker through escape holes in the cap.

2 Setting up the control system. For the typical sprinkler system shown on the preceding page, install three T fittings beyond the vacuum breaker, separating the Ts from the breaker and one another with 2-inch threaded nipples and setting the tapped T inlets downward. Seal the horizontal arm of the outermost T with a tapped plug. Screw a 2-inch nipple into each T and a globe shutoff valve onto each nipple. (Globe valves are more expensive than gate valves, but stand up better under use.) To support the pipes and fittings, fasten wood cleats to the wall behind the horizontal nipples, and clamp the nipples to the cleats.

Installing Pipes and Sprinkler Heads

1 Running the pipe. Working with one circuit at a time (this drawing shows the Number 2 circuit in the plan on pages 118-119), drive stakes at the location of the sprinkler heads as guides to dig trenches for the pipe *(pages 111-112, Steps 2 through 4)*. Assemble all the straight runs of pipe before placing any pipe in the trenches. Measure the distances between the fittings in your plan, and cut pipe to these lengths.

Now install the fittings: an elbow for a sprinkler head at the end of a pipe; a T for a head in a straight run of pipe or for the addition of two pipes; and a T-and-elbow combination, linked by a short spacer, for a head at a corner *(inset)*. Set each completed run of pipe in its trench and join the runs inside the trench. Connect the pipe in the trench to a shutoff valve *(page 113, Step 8)*.

2 Connecting the sprinkler heads. Set a straight stick across the top of the trench over a sprinkler fitting and measure the distance from the stick to the fitting inlet below it *(right)*. This distance is the length of the sprinkler head with its attached riser. Install a threaded adapter on a section of pipe and screw the sprinkler head onto the adapter; then, measuring down from the top of the head, mark and cut the pipe to give the sprinkler-plus-riser length measured. Install the assembly in the fitting, and repeat the procedure for other heads in the system. Remove the sprinkler heads and turn on the water to flush the pipes, then return the heads to the risers for a final test.

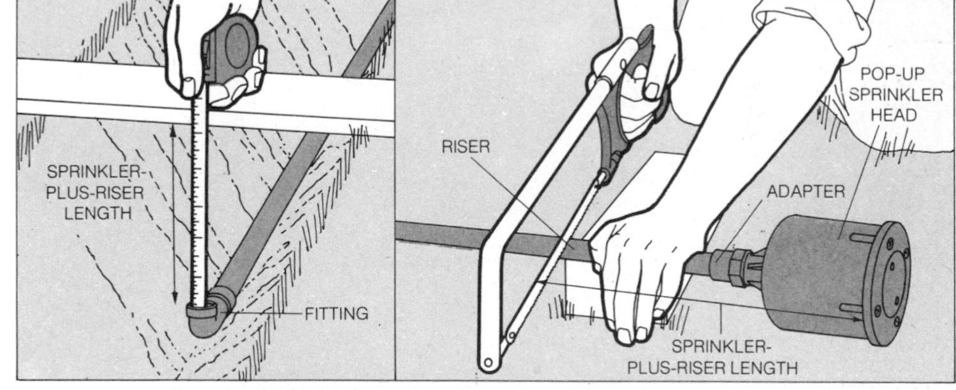

3 Testing the system. With the pipes still exposed in the trenches, turn on the valve for a few minutes to check for leaks in the pipes; at the same time, check each head to be sure that it is producing the spray pattern called for by your plan. If a head sprays an oval pattern rather than a circle, the riser beneath the sprinkler head is not vertical; drive a short vertical stake into the trench next to the riser and tie the riser tightly to the stake with heavy wire, as shown at right. If a head sprays in the wrong direction, simply twist it on the riser to correct the aim. In a few situations, you may have to modify the system by the methods shown on page 122. When you have made all adjustments, fill in the trenches.

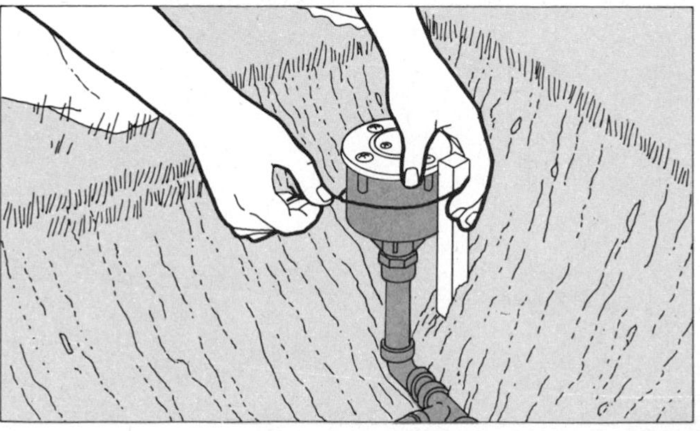

Modifying the System

Usually, a sprinkler system works according to plan, with no more adjustment than straightening a riser or two. In some situations, however, you may have to change spray coverage. Minor variations can be made at the sprinkler heads; major ones involve moving sprinklers.

Changing the aim, or trajectory, of a spray—tilting the head up or down—can increase the reach of the spray. Changing the pattern of distribution makes the spray cover a smaller or greater area. On a simple pop-up head, such changes can be made only by replacing the nozzle. A rotary pop-up, which is more expensive, has separate aim and pattern controls inside its piston.

If these adjustments are inadequate, move the sprinklers for proper coverage. Make the moves with as little additional disturbance to the ground as possible. If you have dug a trench in the wrong place, for example, run short trenches from the existing trench to the new sprinkler positions *(below)*.

Adjusting the spray. To change the pattern or trajectory of the spray from a simple pop-up head *(below)*, replace the nozzle with an alternate model. Pull the piston out of the body and hold it out with the fingers of one hand while you unscrew the nozzle with the other; reverse the procedure to install the new nozzle.

To adjust the type of rotary pop-up normally used for residential areas *(right)*, unscrew the piston cover and reset the controls inside the piston. A plastic disk on a central shaft sets the sprinkler for a full or partial circle of spray. Lift the disk off the shaft, turn it upside down and slip it back onto the shaft to switch from one to the other. Then turn the heavy wire hoops on the rim of the disk to set the width of a partial circle. To change the trajectory of the spray, turn the screw near the side of the piston clockwise or counterclockwise.

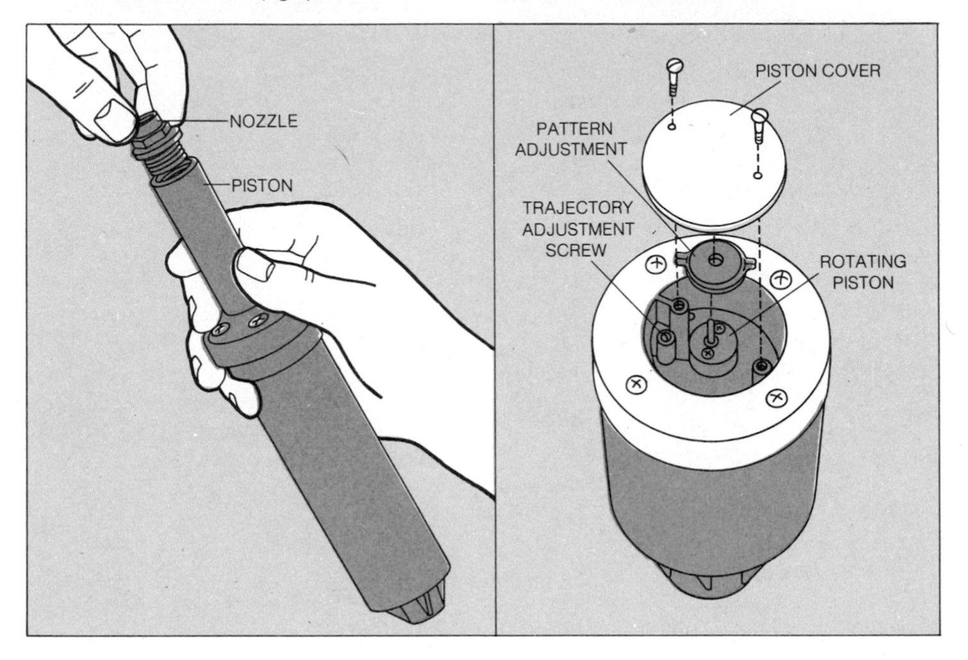

Moving a Set of Sprinkler Heads

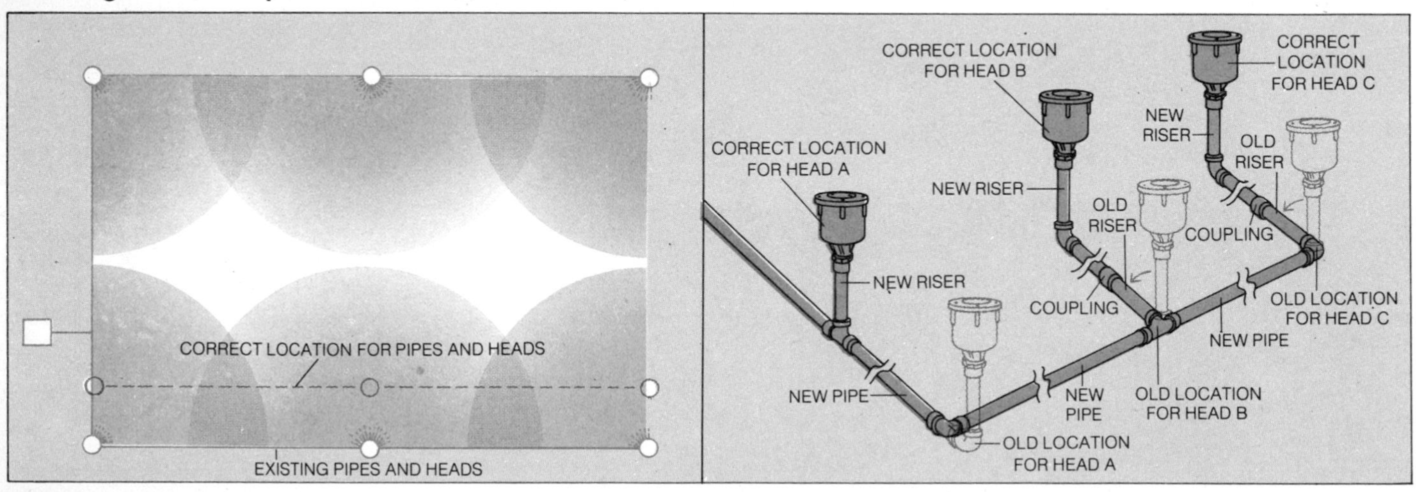

A problem and its solution. In the sprinkler circuit shown above, the heads leave large unwatered patches on portions of the lawn, and no adjustment at the heads can make them cover these patches. Moving the lower pipe and heads to the position of the broken line would correct the coverage—but a much easier solution is shown at right. This pipe-and-sprinkler layout uses the same trench and most of the same pipe. Two short trenches have been dug from the existing trench to the correct positions for sprinklers B and C. The installer cut the existing pipe at the correct position for sprinkler head A, and just beyond the corner T fitting. He cut off and discarded the adapters on the risers of heads B and C, then turned the entire pipe assembly so that the risers lay in the new trenches. Finally, he installed a T and a riser for head A, new pipe and an elbow for the corner, and pipe extensions, elbows and risers for heads B and C.

Automatic Controls

Lawns, shrubs and flowers often need to be watered on different days, and for different lengths of time. Electrical controls can do this job, turning water on and off at the right moments. The owner of the sprinkler system can leave home knowing that his lawn is regularly watered.

Automatic sprinkler controls consist of two parts: a set of electrically controlled valves and a set of electric timers in a control box. Normally, the control box is mounted indoors, the valves are mounted outdoors near the supply lines, and the wires between the two are threaded through a hole in the house wall. The installation on this page, designed for the typical sprinkler system shown on pages 118-119, draws water from an indoor supply line, the usual arrangement in a completely new system. To install automatic controls in a manual system supplied from an outdoor sillcock, replace the manual shutoff valves with electrical valves, then install the control box and make the electrical connections.

1 Control valves for a service line. Drill a hole through the house wall at least 6 inches above the ground. Tap a service line in the basement near the hole *(page 113)*, install a gate shutoff valve on the new pipe and run pipe from the valve through the hole in the wall. Outside the wall, extend the pipe with a 90° elbow and a 2-inch nipple pointed horizontally along the wall. Install a vacuum breaker *(page 120)* on the nipple. Beyond the vacuum breaker, clamp an assembly of three Ts to the wall, linked by 2-inch nipples; seal the outermost T with a tapped plug *(page 120, Step 2)*. Screw a 2-inch nipple into each T inlet and an electrical control valve onto each nipple *(inset)*; the arrow on each valve should point in the direction of water flow. Screw nipples into the valves, and 90° elbows, pointed down, onto the nipples; then run pipe to the sprinkler heads *(page 121)*. If the wall behind the valves is wood, drill a ¼-inch hole to the middle valve and thread the wires from all three valves through the hole. If the wall is not wood, thread the wires through the hole you have made for the main supply pipe.

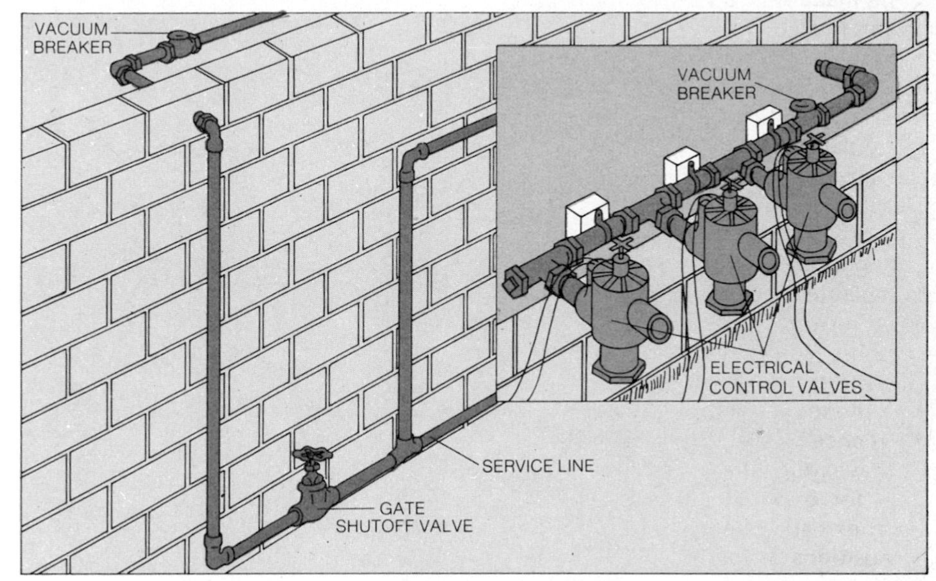

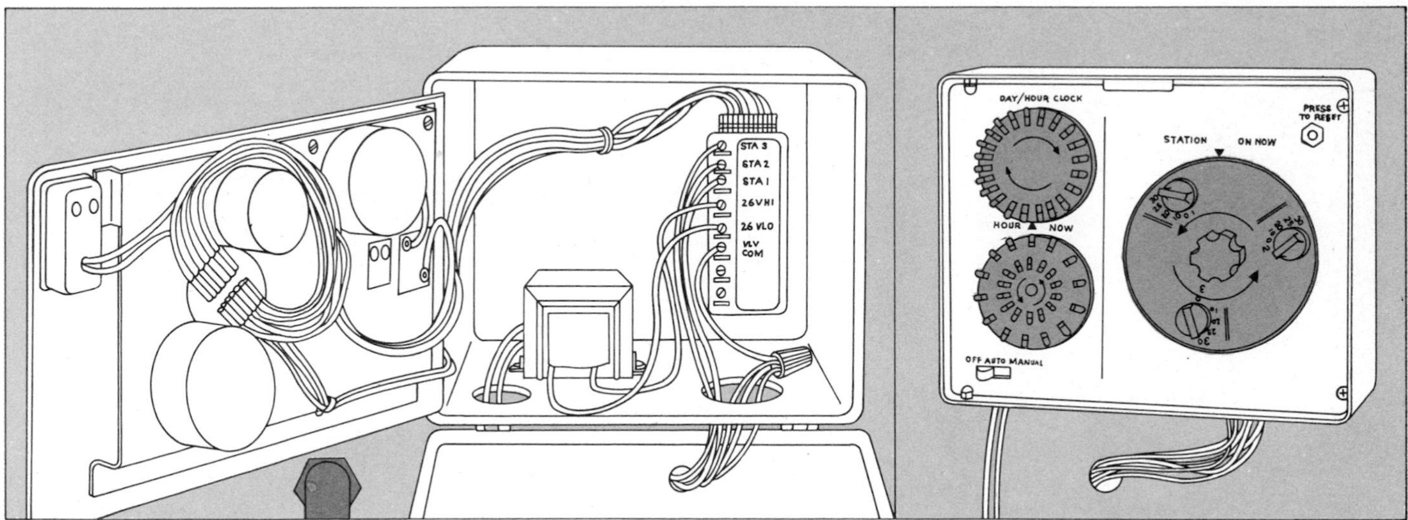

2 Connecting the valves to the control box. Screw the control box to the wall inside the basement near the entrance hole of the valve wires. Swing the front panel of the box open and thread either of the two wires from each valve through the hole directly under the terminals. The numbered station terminals are the control points for the system's circuits; ignore the prewired "26 V" terminals below them. Connect the valve wires to the station terminals, matching the numbers of the valves and the terminals. Thread the remaining valve wires into the box and, using a wire cap, join them to each other and to a short length of wire of the same kind and size; connect this wire to the terminal marked either neutral or valve common *(above, left)*.

Close the front panel, set the "mode of operation" switch to "off" *(above, right)* and plug the control-box wire into an electrical receptacle. The method of setting the controls will vary, but all models have features in common. "Day" and "hour" dials fix the times at which sprinkling begins; individual circuit dials set the length of time each zone is watered. Pushing the mode switch to "automatic" puts the system on automatic control. To override the automatic settings—in a period of heavy rain, for example—push the switch to "manual" and turn the valves on and off as you would in a manual system.

Picture Credits

The sources for the illustrations in this book are shown below. Credits for the pictures from left to right are separated by semicolons, from top to bottom by dashes.

Cover—Ken Kay. 6—Ken Kay. 10,11—Al Freni. 12,13—Drawings by Adolph E. Brotman. 16A through 16H—Ken Kay. 18 through 21—Drawings by Peter McGinn. 22 through 27—Drawings by Adolph E. Brotman. 28—Ken Kay. 30 through 41 —Drawings by Vantage Art, Inc. 42,43 —Drawings by Whitman Studio, Inc. 44 through 47—Drawings by Vantage Art, Inc. 48,49—Drawings by Jim Silks. 50,51 —Drawings by Peter McGinn. 52,53 —Drawings by Whitman Studio, Inc. 54 through 57—Drawings by Peter McGinn. 58 through 67—Drawings by Ray Skibinski. 68—Ken Kay. 70,71—Drawings by Whitman Studio, Inc. 72 through 77 —Drawings by Ray Skibinski. 78,79 —Drawings by Adolph E. Brotman. 80,81 —Drawings by Whitman Studio, Inc. 82 through 89—Drawings by Vantage Art, Inc. 90—Ken Kay. 92,93—Drawings by Adolph E. Brotman. 94 through 97—Drawings by Whitman Studio, Inc. 98 through 101—Drawings by Vantage Art, Inc. 102, 103—Drawings by Whitman Studio, Inc. 104,105—Drawings by Peter McGinn. 106 through 109—Drawings by Ray Skibinski. 110 through 117—Drawings by Nicholas Fasciano. 118,119—Drawings by Nicholas Fasciano, sprinkler design courtesy Oscar Villaverde, Safe-T-Lawn, Inc., Miami, Florida. 120 through 123—Drawings by Nicholas Fasciano.

Acknowledgments

The index/glossary for this book was prepared by Mel Ingber. The editors also wish to thank the following: Edward J. Abele, Richard Deuel, American Standard, Inc., New York, New York, Alsons Corporation, Covina, California; Joseph Carey, Plumber, Rock-Time, New York, New York; Thomas R. Doll, Research Engineer and Thomas M. Jackson, Research Engineer, Davidson Laboratory, Stevens Institute of Technology, Hoboken, New Jersey; Bill Donovan, Powers-Fiat, Plainview, New York; Raymond Durazo, Director, Plastics Pipe Institute, New York, New York; Mr. and Mrs. Warren Hansen, Short Hills, New Jersey; Steve Kelsey, Director of Communication and Merchandising, Moen Division of Stanadyne, Elyria, Ohio; Josam Manufacturing Co., Michigan City, Indiana; Jean Melacon, New York, New York; Steve Pasquini, Northeast District Sales Manager, Safe-T-Lawn, Inc., Carmel, New York; Dr. Stanley Smith, Associate Professor of Electrical Engineering, Stevens Institute of Technology, Hoboken, New Jersey; Steve Solorio, Applications Engineer and Oscar Villaverde, Design Engineer, Safe-T-Lawn, Inc., Miami, Florida; Woodford Manufacturing Co., Des Moines, Iowa.

Index/Glossary

Included in this index are definitions of many of the plumbing terms used in this book. Page references in italics indicate an illustration of the subject mentioned.

Printed in U.S.A.